To Linda

The European Community
Second edition

Allan M. Williams

IBG STUDIES IN GEOGRAPHY

General Editors
Felix Driver and Neil Roberts

IBG Studies in Geography are a range of stimulating texts which critically summarize the latest developments across the entire field of geography. Intended for students around the world, the series is published by Blackwell on behalf of the Institute of British Geographers.

Published

Debt and Development
Stuart Corbridge

Service Industries in the World Economy
Peter Daniels

The Changing Geography in China
Frank Leeming

Critical Issues in Tourism
Gareth Shaw and Allan M. Williams

The European Community, Second edition
Allan M. Williams

In preparation

Geography and Gender
Liz Bondi

Population Geography
A. G. Champion, A. Findlay and E. Graham

Rural Geography
Paul Cloke

The Geography of Crime and Policing
Nick Fyfe

Fluvial Geomorphology
Keith Richards

Russia in the Modern World
Denis Shaw

The Sources and Uses of Energy
John Soussan

Retail Restructuring
Neil Wrigley

THE EUROPEAN COMMUNITY

The Contradictions of Integration
Second edition

Allan M. Williams

BLACKWELL
Oxford UK & Cambridge USA

First published 1991
Second edition published 1994
Reprinted 1995

Blackwell Publishers, the publishing imprint of
Basil Blackwell Ltd.
108 Cowley Road, Oxford, OX4 1JF, UK

Basil Blackwell Inc.
238 Main Street, Cambridge, Massachusetts 02142, USA

British Library Cataloguing in Publication Data
A CIP catalogue record for this book is available from the British Library.

Library of Congress Cataloging-in-Publication Data

Williams, Allan M.
   The European Community: the contradictions of integration
Allan M. Williams. – 2nd ed.
      p.   cm. – (IBG Studies in Geography)
   Includes bibliographical references (p. 239) and index.
   ISBN 0–631–19171–2 (alk. pap.). — ISBN 0–631–19172–0 (pbk.: alk. pap.)
   1. European Economic Community – History.   I. Title.   II. Series.
HC241.2.W475   1994
337.1′42—dc20   93–42149
                                                                CIP

Typeset in 11 on 13 point Plantin by Photoprint, Torquay, Devon
Printed in Great Britain by Athenaeum Press Ltd, Gateshead, Tyne & Wear

This book is printed on acid-free paper.

Contents

List of Figures

List of Tables

Preface

When the series editor, Nigel Thrift, first approached me to write a book on the European Community, I was quietly delighted. Several of my research interests had been converging at EC-level analyses, I was involved in an UNESCO-funded project on Culture and Development in Europe, and my teaching responsibilities were shifting in this direction. After due discussion, contracts were signed and I planned some writing time in the year ahead.

The book I planned to write – and have written – had two main aims. The first of these was to trace the historical evolution of the Community. For this purpose, I have adopted a four-fold periodization of Community development. My interest was not in the minutiae of EC history *per se*. Rather, it was to place in perspective many of the current controversies and debates surrounding the European Community. The 1992 Single Market programme, a Social Charter for Europe, the threat to the environment, the reform of the Common Agricultural Policy are often treated as being 'new' issues which have emerged in the 1980s or 1990s. Yet their roots lie in the very foundations of the Community and they have been central to its subsequent evolution – both conditioning change and being conditioned by other changes. Many of the so-called failures of the EC – such as the farm policy or the neglect of environmental issues – can actually be interpreted as successes in terms of the early goals of and constraints on the Community. All these issues can only be fully appreciated in context of how the Community was founded, and rapidly – if partially – integrated in the 1960s, before stagnating politically and economically in the 1970s and early 1980s.

The second aim of the book was to examine some of the relationships between the global and the intra-EC levels. Neither of these offers a complete explanation of how the Community has evolved. Instead, it is important that both perspectives are brought together. Many of the major impetuses to Community development have been global events. The East–West Cold War, and the USA vs. Japan vs. EC economic struggle are only the most obvious of these. However, the way in which such mega-trends impact on the Community in detail are conditioned by the internal structure of the Community. The balance between the EC institutions, the different interests of the member states and, above all, their domestic policies are all central to this. In the post-1945 period this would have been subject to universalization tendencies at the economic, political and social levels but in Western Europe these have been resolved into two parallel if linked processes: Europeanization and globalization. Both have moulded the way in which the Community has developed.

In this book I have tried to show how these various levels of processes have influenced the European Community space. Integration has advanced to the point where, in many respects, we can talk of an EC economic space. Essentially, the 1992 programme is about deepening this space. However, it is an uneven economic space, in that some countries and some regions are far more dominant than others. The EC economic space is also not matched by either an EC political space or an EC social space. These are still far more fragmented into national spaces and lack Community-wide application. Not least decision-making is still founded on inter-governmentalism rather than federalism.

What I had not bargained for when, early in 1989, I agreed to write this book, was the momentous events that were to unfold in Eastern Europe later that same year. The slow trickle of economic and political thaw suddenly turned into a flood of reforms. This culminated in the horrific, yet optimistic, scenes in Romania in December, which transfixed western television audiences over the Christmas period. Even then, the process of opening between Western and Eastern Europe was far from complete. For example, while this preface was being written, the date for German unification was announced. The political and economic transformation of Eastern Europe is likely to be a long process and, in many respects, may not be completed until the next century. However, these remarkable changes have necessitated a major rethink of what constitutes the European space. Depending on our definitions of Europe, this extends from Lisbon to the German/Polish border, the Soviet border or the Ural mountains. Each of these has very different implications for the way in which the EC space may expand in future. I have tried to incorporate some of these concerns into this volume but, of course, cannot claim access to a crystal ball which will foretell events in the 1990s.

Finally, I wish to thank three individuals for their help with this book. Nigel Thrift for the encouragement to prepare the manuscript, Tracy Reeves for helping with word-processing and Terry Bacon for his imaginative cartography. Their assistance was invaluable but, as usual, the responsibility for any errors or misinterpretations rests with the author.

Allan M. Williams

Preface to the Second Edition

It is always gratifying to be asked to write a second edition of a book, and the early invitation to do so in this case is indicative of the speed at which events have unfolded on the European stage. This, of course, brings its own challenges, given that any book must necessarily cut across long-term evolutionary trends at one terminal point in time. In the case of the European Community there have been particularly sharp oscillations in mood, decision-taking and events with respect to both individual policies and macro-scale integration. It is no exaggeration to say that any attempt to sum up the present state of the Community and of European integration would come to very different conclusions depending on whether it was written in 1991, 1992 or 1993. This was recognized in the preface to the first edition, written at the time of German unification and the opening of Eastern and Central Europe. Since then we have seen the emergence of the true costs of German unification and of the difficulties of reintegrating East and West Europe, as well as the problematic passage of the Maastricht agreement and of Economic and Monetary Union, not to mention the depressing tragedy being played out in Yugoslavia. Each of these has presented sharp lessons about the role that the Community is able to play in larger settings, as well as the limitations of European integration. The speed and the scale of events does not excuse us from trying to interpret events. Instead they demand that we be more sensitive to the limitations of our own time-bound perspectives, and more catholic in our analytical frameworks. One of the aims of this book has been to show that geographers have a role to play in such analyses alongside other social scientists.

Another difficulty in writing about the Community is that of changing terminology. An early example was the change in title from European Economic Community to European Communities, while in 1993 the EC became parts of the umbrella EU after the enactment of the Treaty of European Union. There was a considerable time lag in popular and even academic acceptance of the first change in nomenclature. This, together with the fact that this manuscript was completed before the Maastricht agreement had received its final approval in the German courts, has led to a decision not to use the term European Union in either the title of this volume or as a general term of reference in the text.

Finally, I wish to acknowledge the assistance of the following: John Davey for encouraging me to prepare this second edition; Terry Bacon for preparing the new and revised maps; and Jo Small for typing all the tables and references which were beyond my own word-processing skills.

Allan M. Williams

ONE

Growth, Global Interdependence and Contradictions

The emergence, growth and expansion of the European Community (EC) has been the outstanding feature of the economic and political geography of Europe since the 1950s. It is a complex phenomenon – both politically and economically – and can be studied at a number of levels. Firstly, there is the immediate economic and political context of post-war Europe, shaped by the experiences of the Second World War and by the East–West partition of Europe. The growth of the EC played a part in the economic recovery of Western Europe as well as being fostered by it. Secondly, the form and functions of the EC were shaped by the global competition between the superpowers. Initially, this was the global military and political conflict between the USA and the USSR. By the 1980s it had become a three-cornered economic contest between the USA, Japan and Western Europe set against the geopolitical uncertainties which followed the demise of the Soviet Union. Thirdly, the growth and integration of the EC has followed an uneven and often troubled course. Not least, there has been constant tension between the powers and responsibilities of the EC and of its constituent member states. This has been compounded by the struggle for power between different institutions within the EC: the European Parliament, Commission and Council of Ministers have been locked in constant political contest. Despite these difficulties, the EC has recorded some notable economic achievements, in terms of growth and policy-making. Yet economic growth has engendered a number of problems for the Community; not least, contradictions have emerged between economic, social, regional and environmental goals. Their resolution has further exasperated the tensions – more often destructive

1

than creative – between individual states and the EC. Persistent economic recession, throughout much of the 1980s and 1990s, has also impoverished the environment for European integration.

The aim of this book is to analyse the aims, operations and policies of the EC, and their consequences, in relation to individual states, Europe and the global political economy. As the role of the EC on the world stage, its policies and its composition have changed significantly over time, such an analysis requires an historical perspective. This is provided through a simple periodization corresponding to the main phases in the internal organization and the external role of the EC. Chapter Two looks at the origins of the EC, set in context of post-war reconstruction and the USA's global hegemony. Chapter Three looks at the formative years, from the late 1950s to the early 1970s, when the major policy framework was established during a period of strong global economic growth. By the 1970s the global economy was entering a period of intensified economic competition and more uncertain growth: chapter Four examines the economic stagnation of the EC during these years when the dynamism of integration seemed exhausted. By the 1980s the Community had reached a crisis point, beset by both external constraints and internal conflicts. Chapter Five considers the responses to these difficulties, especially the origins, aims and consequences of the Single Market programme. Chapter Six looks at the further advancement of European integration, via the Maastricht Agreement, as well as the difficulties that this has encountered. Finally, chapter Seven examines the state of the EC in the 1980s. A number of contradictions are identified, and the EC's responses are evaluated.

A central theme is the making and the remaking of the European space. The Cold War led to a partition of the European political and economic space, and within Western Europe the EC and EFTA created new forms of organizations. In time, these became new realities in the European space and they remained virtually unchallenged until the late 1980s. In recent years, the Single Market programme, Maastricht and the East–West political opening have changed the parameters which govern the European space. The end-product of the remaking of this space is as yet uncertain but clearly has involved the European Economic Space (the EC and EFTA) and could extend to the European Homeland (stretching from the Urals to the Western Atlantic borders of Europe).

The European Context

The European Community came into existence on 1 January 1958. At that stage it was known as the European Economic Community and it existed alongside EURATOM and the European Coal and Steel Community (which had been established six years earlier). These were later amalgamated into the single European Community and this is the term which is normally employed in this book. Although based on earlier European experiments in international co-operation, such as the Benelux Customs Union, the EC marked a significant change in the organization of the European economy. Its membership involved two of the three major economic powers within Europe (France and the Federal Republic of Germany) and its aims – as stated in the Treaty of Rome – encompassed social and political as well as economic reorganization. Set against the history of inter-state conflict in Europe this reorganization was a genuinely revolutionary initiative. As Lee (1990, 236–7) argues: 'The non-congruence of European geography, society and state remains a major motor of European history in which the contemporary quest for a Europe constituted by institutions is a very recent manifestation.' Although not fully appreciated at the time, this new institutional arrangement would eventually shape the political and economic geography not only of Western Europe but of the world.

Two issues were particularly important in the background to the signing of the Treaty of Rome. Firstly, the need for the political reconstruction of Western Europe, in the context of the East–West Cold War. The Federal Republic of Germany had to be reconciled with its neighbours and wartime opponents, especially France. Secondly, there was the powerful logic of capital accumulation. There had been rapid economic recovery after the end of the Second

World War, and the 1950s had witnessed the beginnings of what was to prove a long and sustained boom in consumption and investment. However, it also exposed the critical need for economic reorganization. Production was becoming increasingly globalized and required the assemblage of capital and labour at the international scale. The logics of market access and scale economies also pointed to the need for transnational strategies in both sales and production. Therefore, the creation of a new European framework offered *one* response to the needs of private capital in Europe. In addition, Europe's post-1945 growth rested largely on the economic hegemony of the United States. This served to underline the economic weakness of individual European states on the world stage, whether in political or economic terms, and the need to foster pan-Europeanism.

Although the Treaty of Rome is a long and complex document, in scope if not in detail, its short-term objectives were simple: creation of a customs union (without internal tariffs or barriers to trade), a common external tariff, and common markets for capital and labour. There was also a second economic agenda for the development of Community-wide policy frameworks for industry, agriculture, transport and so on. In practice, early progress was gradualist with the emphasis on practical, if piecemeal, achievements. For example, the target dates for achieving the customs union were postponed until the late 1960s. Even so, in setting up the EC an institutional reform had been initiated which, eventually, would shape the whole process of capital accumulation in the European arena. Not only did it influence trade and investment in Western Europe but, arguably, it set in train processes which made the further enlargement of the EC inevitable.

Global conditions could hardly have been more favourable for the foundation of the EC. The Community was born just as the post-war boom was gaining momentum, and as Western Europe was recording growth and investment rates without parallel in the first half of the twentieth century. For example, between 1950 and 1970 Western Europe experienced growth rates of 4.4 per cent per annum, which considerably surpassed global economic growth rates (Aldcroft 1980). As a result, the context was favourable for the EC to benefit from the twin fruits of economic integration: trade diversion (from other countries) and trade creation (resulting from market enlargement). Furthermore, potentially contentious trade diversion (from the rest of Europe and the USA) was made less painful and contentious as there was strong global expansion of trade and investment in most countries. This was in marked contrast to the 1980s and 1990s, when global recession has sharpened the increasingly sensitive trade relationships between the EC and the rest of the world, especially the USA.

The EC – along with the remainder of Western Europe – prospered during the 1950s and 1960s and the initial post-war trade deficit with the USA had turned into a European surplus as early as the late 1950s (Ballance and Sinclair 1983). This helped to make the 1960s a decade of impressive achievement for the EC. The customs union was completed in 1968, well ahead of schedule, and the frameworks of major policies such as the Common Agricultural Policy and the Social Fund were established. After several false starts an agreement was also signed in 1972 for the accession of three new members to the Community: the UK, Ireland and Denmark. In the same year the EC reached the high point of its early achievements. At the 1972 Paris summit the member states agreed on the need for greater economic and monetary integration by the end of the decade. Economic union appeared to be in sight.

This was to prove a false dawn. In reality, the global economy was already faltering by the early 1970s, and the 1974 and 1979 oil crises gave it a firm downward push, leading to world recession in the 1980s. The very growth of the EC and of its share of world trade had, of course, contributed to a decline in the hegemony of the USA. This, in turn, had led to chronic American trade deficits and dwindling gold reserves which contributed to the introduction of flexible exchange rates in 1973, ushering in an era of international economic instability. In this colder economic climate the Community lost much of its positive dynamism and, instead, foundered on indecision, political obstructionism and a profound failure to implement essential internal reforms. It seemed unable to respond in a positive fashion to either the growing economic challenge of

Japan or to the needs of Western Europe as a whole. For example, the 1977 applications of Portugal and Spain to join the EC were not approved until 1985 even though, after emerging from the Salazar and Franco dictatorships in 1974 and 1975 respectively, these countries had critical need of the economic and political framework provided by EC membership.

It would, however, be misleading to see the difficulties of the EC as simply a function of global downturn in the 1970s and 1980s. To a large degree the very conception of the EC was an economic abstraction. The dissection of the remnants of the Austro-Hungarian empire at the close of the First World War had already led to the fragmentation of one previously unified economic zone. Protectionism had further disrupted European economic relationships during the 1930s (Pollard 1981). The Cold War and economic partition was only one more, if critical, step in this economic division of Europe. Nowhere was this more deeply felt than in FR Germany, where parts of the pre-war national economy, as well as traditional trading partners (such as Czechoslovakia) ended up on different sides of the great political, and increasingly economic divide, the Iron Curtain.

In some ways, the establishment of the EC served to increase European fragmentation. For a variety of complex reasons there were only six initial members of the Community (see Blacksell 1981). Austria and Finland, by virtue of the conditions of the post-war settlement, were precluded from membership. Spain and Portugal were excluded by their authoritarian governments, while Sweden and Switzerland were committed to the politics of neutrality. The UK turned its back on the EC, convinced it had a different role as the head of the Commonwealth and, indeed, as an independent superpower. As a result, the EC in 1958 occupied only one fragment of Western European economic and political space, let alone of the larger Europe.

As the EC cut across so many traditional economic relationships within Europe, it contributed to the creation of a new economic geography of the continent. Trade-diversion effects were considerable, not least because many of the EC member states had previously had much stronger ties with non-members than with fellow-members. This was particularly true of FR Germany which occupied a pivotal role in the European economy as a whole, not just within the Community of Six. New patterns of trade and investment emerged within the EC, which had profound implications for the rest of Western Europe. Partly as a defensive stratagem, most of the remaining Western European economies grouped themselves around the UK in the rival European Free Trade Association during the 1960s. However, the UK economy was not strong enough to sustain this group, especially in the face of the weight and dynamism of capital accumulation in the EC. Not surprisingly, EFTA as a UK-centred association did not survive for long. In January 1973 three of its members left EFTA and joined the EC (see figure 1.1). When Greece, Spain and Portugal emerged from dictatorial governments in 1974–5 there was no question as to which European organization they wanted to join. Hence the process of EC enlargement continued during the 1980s (see figure 1.1).

By the mid-1980s the EC had 88 per cent of the GDP and 90 per cent of the population of Western Europe. Therefore, when the Community finally set aside indecisiveness and agreed to the 1992 Single Market programme and the 1986 Single European Act, this provided a catalyst for changes in the sub-continent as a whole. It led the EFTA group to negotiate the creation of 'The European Economic Space'. This was a proposed extension of the customs union and the common market to virtually the whole of Western Europe. By 1990, even 'the 1992 programme' had been over-shadowed by momentous events in Eastern Europe. Democratic reforms spread through the region, accompanied by moves to privatization and to free markets. It is significant that the EC played a central role in the economic and political dialogue between West and East. In some ways its new relationship to Eastern Europe, and its commitment to the economic rebuilding of the region, was similar to the role that the USA and Marshall Aid had played in the reconstruction of Western Europe after 1945. There could be no more poignant comment on the status of the EC, not only at the European scale but also on the world stage. With barriers to East–

a) Membership and enlargement

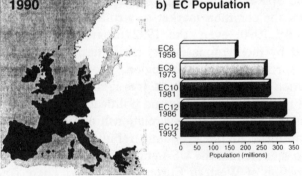

Figure 1.1   The European Community

West trade and investment tumbling, the EC was poised, once again, to play a central role in the reshaping of the political and economic geography of Europe. The most obvious instance of this was the effective fourth enlargement of the Community following German unification in October 1990 (figure 1.1).

The Global Context

The EC was established in a period of unsurpassed economic growth in Western Europe in the 1950s and 1960s. Most of the future members of the EC had already recovered their pre-war levels of production and investment by 1950. When the Treaty of Rome was signed in 1957, Western Europe had benefited from almost a decade of strong growth and rising levels of exports, both within Europe and with other major world trading blocs. Indeed, by the late 1950s Western Europe was moving from trade deficits with the USA to structural trade surpluses. Thereafter, the 1960s were years of strong global economic expansion which provided a favourable backcloth for the establishment of the Community, its institutions and policies.

Growing global economic interdependence was another characteristic of the post-war years. While the EC benefited from this during its first decade and a half of existence, it also meant that the Community was not immune to the global crises of the 1970s and 1980s. In part, the very growth of the EC contributed to these crises, for its structural trade surpluses were one factor in the declining economic hegemony of the USA. In turn, the relative weakening of the USA led to a collapse of the monetary system, which disrupted trade and investment. There were, of course, other reasons for the economic crises of the 1970s and the 1980s. In particular, competition from Japan and the Newly Industrializing Countries (NICs) contributed to a crisis of overproduction, while the oil price shocks of 1973–4 and 1979 also depressed Western economies (Williams 1987).

Whatever the causes of the global crisis, the system of global interdependence meant that the EC could not escape its consequences. Indeed, the EC suffered more from reduced growth, inflation and high unemployment in the 1980s than did its major rivals, the USA and Japan. Furthermore, by the early 1980s it was evident that the EC was failing to compete globally in a number of key economic sectors, especially those involving advanced technology. Even FR Germany, which had emerged as the dominant economy in the EC, could not compete with Japanese and USA multinationals in such strategic industries as computer manufacture, consumer electronics and robotics. Both the EC and the member states sought to respond to these challenges, frequently resorting to protectionism. However, the nature of global interdependence – seen in the threat of international trade wars and in the growth of American and Japanese multinationals operating *within* the EC – severely limited the scope for protectionism. The search for more positive policy responses led to the evolution of EC technology policies. It also contributed to moves to reform the EC budget, so as to reduce expenditure on the CAP in order to release resources for other policy objectives. More significantly, awareness of the competitive weaknesses of the EC within the global economic system provided a major input to the amalgam of political and economic forces which was to lead to the 1992 Single Market programme. This gives credence to Ross's (1991, 49) view that the EC 'has thus far been constructed out of the materials of capitalist crisis, by gifted politicians'.

The emerging global political system also shaped the evolution of the EC. Most importantly, the global struggle between the two superpowers – the USA and the USSR – was mirrored in the East–West partition of Europe. The Iron Curtain became the dominant influence on the political evolution of post-war Europe. Most of Western and Eastern Europe – with a few exceptions such as Finland and Yugoslavia (see Blacksell 1981) – became grouped into competing military alliances, NATO and the Warsaw Pact, on either side of the Iron Curtain. While essentially a political and military divide this also had economic significance. For example, by the late 1980s only about 5 per cent of the trade of the member states was with Eastern Europe. To some extent this was due to the trade diversion effects which followed the creation of the EC and, in Eastern Europe, of COMECON. However, more fundamental were the difficulties of generating trade between market and centrally planned economies.

The political division of Europe also influenced the membership of the Community. The EC, at its inception, was seen as an important economic and political ally of the USA. Although there was later to be economic conflict between these two trading blocs, the accession of the UK to the EC in the early 1970s meant that the political ties of the USA with the Community were further strengthened. As a result, those Western European states which chose, or had, to remain internationally neutral – such as Sweden, Switzerland, Finland and Austria – could not contemplate membership of the EC. The thaw in East–West relations in the late 1980s and the political revolution in Eastern Europe in 1989–90 has, however, radically modified the international political context and redefined notions of neutrality.

While the EC was conceived of as both a political and an economic entity, it has been much more active in international economic affairs than in political ones. Most of Western Europe's political influence has developed at the level of individual states, with France and the UK, given their permanent representation on the Security Council of the United Nations, being particularly important. At the associational level, the American-dominated NATO has been the main outlet for Western European foreign policy. Therefore, until recently, the EC had a limited role in international foreign policy. However, this is gradually changing. The EC has sought to establish common foreign policy positions with respect to a number of issues, including relationships with South Africa and the Middle East. While their success has been limited, the 1970 Luxembourg Report establishing European Political Co-operation (EPC) was of greater significance. The attempt to secure greater political union in the 1980s and the 1990s has contributed to the evolution of an international political role for the EC. This has been facilitated by the way that the Community has emerged as a potentially critical actor in the economic and political reorganization of a new Europe, following the opening of Eastern Europe in 1989–90. It is, for example, the principal co-ordinator of the European Bank for Reconstruction and Development. However, the history of the crisis in what was once Yugoslavia has provided a poignant lesson in the limitations of EC foreign-policy interventions.

Integration and Contradictions

During its relatively brief existence – little more than thirty years – the EC has achieved some remarkable successes. Common policies have been formulated, a pan-European electoral system has been established and there have been consistently high economic growth rates. Moreover, the EC has been recognized as an important economic player not only on the European stage but also at a global scale. This is certainly true of economic relationships and the EC has taken over many of the functions previously exercised by individual states, such as GATT negotiations. It is also increasingly true of political relationships, as the EC edges its way towards a common foreign policy and implementing European Political Co-operation. However, despite these significant advances, progress in the EC has been marked by a series of internal and external crises. There are a number of inherent contradictions in the EC, both in its institutions and its policies, and these have continued to pose obstacles to further progress.

The first, and in some ways the most important, set of contradictions centre on the institutions of the EC and its system of governance. The political organization of the EC is unique although, in essence, it is a form of intergovernmentalism. Power is divided between three bodies: the European Commission, the Council of Ministers (heads of national governments) and the European Parliament. The balance of power has shifted over time but, ultimately, has largely rested with the Council of Ministers, with the Commission having a secondary role. While the Commission initiates policies and may challenge the actions of individual states, taking them to the European Court if necessary, ultimately its actions and its finances depend on the approval of the Council of Ministers. The European Parliament has a more limited role to play and its only real power lies in its right to approve the budget. Its powers were substantially increased by the 1986 Single European Act, but they still only allow it to delay legislation or to require unanimous voting in the Council of Ministers to overturn a parliamentary vote. These institutional arrangements have generated the claim that there is a democratic deficit in the Community.

The 1966 Luxembourg compromise, which established the need for unanimous voting, gave additional importance to the role of the Council of Ministers. This is problematic because, at heart, the EC is an economic institution and most of its policies are concerned with resource allocation. Therefore, the Luxembourg compromise laid the EC open to the requirement of satisfying the national aspirations of *all* its member states. Decision-making was further complicated in the 1970s and 1980s by enlargement of the Community from six to twelve members, and by a growing resource squeeze in the face of mounting national and sectoral demands. Not least, the failures to resolve the UK's demands for a budgetary rebate and to reform the Common Agricultural Policy (CAP) bedevilled all attempts to advance integration and to develop common policies in this period. This exposed the limitations of intergovernmentalism, that is of joint decision-making by the member states rather than by a single federal body. By the early 1980s decision-making in the EC was characterized by minimalism. On most occasions minimal progress was made in response to minimal concessions by the heads of member states. The 1986 Single European Act served to weaken, if not to break, this stasis in decision-making. However, the EC continues to be characterized by intergovernmentalism. It had also failed to achieve the political union or federalism that some member states had hoped for. But the Single European Act was informed by what became known as Delors's 'Russian doll' strategy, for it contained the seeds of further economic, monetary and political integration, which were eventually brought to fruition in the Maastricht agreement.

A second set of contradictions centred on the balance between growth and distributional aims in the EC. The Treaty of Rome does emphasize the need to ameliorate gross inequalities within the Community but, in practice, its policies have been orientated towards economic growth. As a result, there have been a number of major contradictions between the different aims and the policies of the EC:

Monopoly capital.   The creation of a single market tends to favour large-scale capital. Yet this may lead to oligopoly or even monopoly powers which are, of course, contrary to the EC's avowed aims of fostering competition.

Social inequality.   Economic growth has been accompanied by the continuance or deepening of social inequalities within the EC. As these are endemic to capitalist societies, there are limitations to the extent to which the EC – any more than national governments – can significantly modify such inequalities. The major EC response to the existence of a European underclass has been the Social Fund, which has targeted assistance at some of the more economically disadvantaged groups in society, such as migrants, the long-term unemployed and handicapped persons.

Regional inequality.   Regional inequalities have been widespread within the EC since its foundation. Subsequent enlargements have only served to widen these differences. The European Regional Development Fund has been created to try and ameliorate such spatial inequalities, but its successes have been limited. Again, this is largely because regional inequalities are endemic to capitalist economies. Indeed, they are an essential prerequisite of continued capital accumulation.

The environment.   The single-minded pursuit of economic goals has inevitably brought the policies of the EC into conflict with environmental objectives. The fact that so many environmental problems are international in nature has highlighted the potentially important role that the EC has to play in this area. However, as in other policy areas, the EC is beset by the need to balance sectoral interests and by the limitations of intergovernmentalism as a form of decision-making.

The North–South divide. As one of the three economic superpowers, and the historic heart of nineteenth- and twentieth-century imperialism, the Community has particular responsibilities to the Third World. In the area of trade, it has sought to formalize these links in the Lomé conventions. But there are still unresolved contradictions between the trade and aid needs of the Third World, and the economic interests of the member states, as well as between domestic politics and the needs of refugees and international labour migrants.

All these contradictions potentially will be deepened by the 1992 Single Market programme with its emphasis on larger markets, productivity and competitiveness. Furthermore, the capacity of the EC to respond to these challenges is conditioned by the need to remain competitive with the USA and Japan. In other words, while the EC remains essentially a champion of private capital, and while it is locked into a system of global economic interdependence, the scope for redistributional policies or for constraints on the actions of capital will be limited. This is further exaggerated by the need of the EC to evolve policies which meet the requirements of member states with often diverse interests. For example, the Southern and Northern member states have very different roles in the international division of labour, both within Europe and at the global scale. Their priorities across the whole range of social, economic and environmental policies are likely to be very different. EC common policies need to take into account not only these divisions, but also the interests of the Community itself, as an increasingly independent body.

TWO

From the Ashes: the Creation of the European Community

Europe in Ruins

The inter-war years had been a difficult period for Europe. The peace settlement at the end of the First World War had dismembered the remains of the Austro-Hungarian empire, causing severe economic disruption. Once well-integrated industries were separated by new state boundaries as the political map of Europe was redrawn (see figure 2.1). For example, the textile industry was fragmented with spinning mills and weaving factories being divided between Austria and Czechoslovakia. Germany was similarly affected by the new political geography, especially by the dislocation in its heavy industries following the transfer of the Lorraine iron ore fields to France. These specifically European difficulties were compounded by the global recession of the 1930s. In the face of a world crisis of overproduction and rising unemployment, most developed economies retreated into protectionism which, in the medium term, only added to the difficulties of several segments of export-orientated capital.

An important element in the economic recovery of Europe in the late 1930s was remilitarization in response to the threatened expansionism of Hitler's Third Reich. In the longer term, however, the Second World War resulted in economic catastrophe for most of Europe. This can not be fully quantified but a few indicators suggest the extent of the economic damage. There had been extensive destruction of capital stock, estimated at a net reduction of 6–9 per cent in France and the Netherlands (Aldcroft 1980), which must be set against even the modest cumulative increases that would

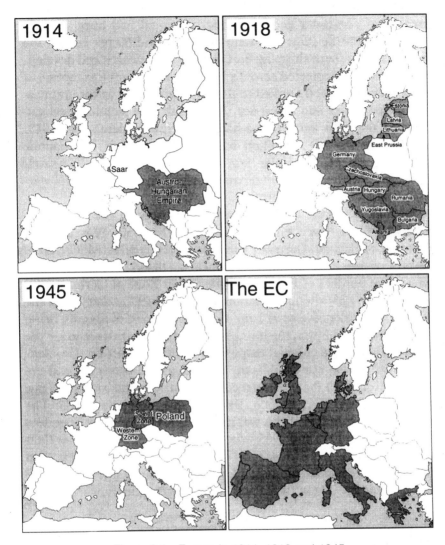

Figure 2.1   Europe in 1914, 1918 and 1945

otherwise have been expected. The stock of net *industrial* capital was largely unchanged, although industrial growth had been highly uneven, with expansion limited to the war-related sectors at the expense of consumer goods. Transport capital had been severely damaged with the railway system disrupted and 60 per cent of the European merchant shipping fleet destroyed. Research and development were also neglected across a broad front, with the exception of a few sections directly linked to the war effort, such as communications, electronics and aerospace. As a result, the technological gap between the USA and Europe had widened considerably by 1945. Finally, the war left 20 million dead in Europe which, in economic terms, represented a major loss of skilled labour. Only neutral Portugal, Sweden and Switzerland escaped the worst of this wartime destruction and disruption.

Further economic disruption awaited Europe at the end of the Second World War. Firstly, there was the Cold War and the partition of Europe. Germany was most severely affected with large parts of its pre-1938 territories disappearing behind the Iron Curtain within the borders of Poland and the Soviet Zone of Germany (see figure 2.1). This disrupted labour and product markets with the factory networks of several large German companies literally being torn apart. In addition, the immediate post war years were to see the dismantling of the colonial empires of Northern Europe. Decolonization in Africa, the Caribbean, the Indian sub-continent and the Far East removed the protected colonial markets and sources of raw materials of the UK, France, Belgium and the Netherlands. The most potent indicator of this was the creation of 57 new independent states between 1945 and 1965, whose formal dependence, if not their actual economic dependence, on Western Europe was far less than that of the previous colonies. However, although Europe in 1945 was in disarray it was to recover surprisingly quickly.

Europe and the New International Economic Order

Western Europe was able to recover rapidly from the economic catastrophes of the Second World War because it benefited from the creation of a new global economic order based on American hegemony, trade liberalization, a stable multilateral exchange system and the boost of Marshall Aid in 1947–8. The key was probably the economic hegemony of the USA. In part this stemmed from the weakness of Europe which allowed the USA to extend its relative economic lead. In addition, the American economy had benefited considerably from the Second World War, without suffering the destruction that Europe had experienced. The main advantages which accrued to the USA were the accumulation of gold reserves (partly as the result of European purchases of war materials), the opportunities to break into Third World markets previously dominated by European companies, the extension of its technological lead and the war-time expansion of its economy (see Williams 1987, chapter One).

A dominant American economy provided the leadership necessary to ensure that some of the economic disasters of the 1920s and 1930s – protectionism and global recession – were avoided, or at least postponed. The Bretton Woods agreements of 1945 provided essential international monetary reform: the dollar and the pound became reserve currencies, with the former having a fixed gold conversion rate. This provided a stable international exchange system which permitted a rapid return to a multilateral payments basis for international trade. There were also parallel trade liberalization reforms, mainly via the General Agreement on Tariffs and Trade (GATT). Beginning in 1947, several rounds of GATT negotiations secured extensive trade liberalization in the post-war years. Although these have by no means eliminated all trade disputes – for example, witness the 'pasta' war between the USA and the EC in the 1980s – they have provided a framework for the continued internationalization of the economies of Western Europe. The main limitation of the GATT negotiations has been their overwhelming focus on manufactured goods, and the virtual exclusion of agriculture and services.

Despite the establishment of a new international economic framework, global – and, therefore, European – economic recovery was slow to take off in the period 1945–9. There were a number of reasons for this. The pound was too weak to act as an international reserve currency and this led to the destabilizing 1947 sterling crisis.

More importantly, the German economy, which was pivotal to the European economy as a whole, was slow to recover under the burden of reparations payments to its wartime opponents. Finally, the winter of 1947 was exceptionally severe and led to widespread energy shortfalls. Capital accumulation in Western Europe was still struggling to recover three years after the end of hostilities.

The American response to these economic difficulties – and to the perceived threat of growing support for Western European communist parties – was the introduction of the Marshall Aid Economic Recovery Programme. This symbolized the high level of interdependence in the global economy. Expansion of the American economy was being held back by Europe's economic stagnation. The USA had a massive trade surplus with Europe which, consequently, experienced a dollar shortage and a structural constraint on the ability to expand demand for American goods. The Marshall Aid programme sought to break this vicious circle by making $13 billion available to Europe in grants and soft loans. Three-quarters of the funds were granted to just five countries: France, Italy, the Netherlands, West Germany and the UK. The first four of these later came together to form the core of the European Community, which emphasizes the key role that the USA played in supporting its creation. In the short term, Marshall Aid paid for essential Western European imports of raw materials and technology, while in the longer term it encouraged acceptance of trade liberalization and multilateral trade payments.

By 1949 European recovery seemed to be well-established and, indeed, the next twenty years were characterized by high and sustained economic growth rates. The reasons for this were complex and included the internationalization of economic activity within the frameworks provided by GATT and Bretton Woods. In addition, there was a high level of demand throughout the 1950s and 1960s fuelled, at first, by post-war reconstruction and later by a boom in consumer durable goods. On the supply side there was the favourable combination of high investment rates, major technological advances in industries such as electronics and petrochemicals, and the availability of a large supply of low cost labour. This long period of economic expansion was to provide highly favourable conditions for the establishment of new forms of economic and political association in Western Europe. While this would lead eventually to the formation of the European Community, there was no inevitability in such an outcome, as will be seen in the next section.

The Agenda for European Integration

At the end of the Second World War there was a climate of opinion in much of Europe which favoured the creation of new forms of political association. For some this was a defensive measure to limit the possibilities of major wars being fought on European soil in the future. But for political idealists, such as Spinelli of Italy and Brugmans of the Netherlands, European federalism was a positive goal in its own right. Several notable speeches and books marked the debate on the desirability and feasibility of European Union, but the most influential was by that most pragmatic of statesmen, Winston Churchill. At Zurich, in September 1946, he appealed for a new spirit of co-operation in Europe:

> The first step in the recreation of the European family must be a partnership between France and Germany. In this way only can France recover the moral leadership of Europe. There can be no revival of Europe without a spiritually great France and a spiritually great Germany. . . . In all this urgent work, France and Germany must take the lead together.

This speech had two notable features. Firstly, it accurately pinpointed the critical relationship between France and Germany which, in due course, would be the nexus for the creation and evolution of the European Community. Secondly, Churchill did not see a role for the UK in any new European association. Instead, the UK would have its own independent role as the head of the Commonwealth and as the special ally of the USA. The UK's global, rather than European, aspirations would profoundly influence the eventual emergence and subsequent evolution of the European Community.

Global politics were also pointing to the need for European integration. The new global balance of political and military power between the USA and the USSR was likely to relegate individual Western Europe states to minor roles in world affairs. Indeed, at a later date Walter Hallstein (1972) was to argue that greater European union had become necessary 'to bring to an end the demoralising situation in which Europe was at the mercy of political decisions taken by others outside Europe'. In addition, the Cold War and the lowering of the Iron Curtain – especially after the outbreak of the Korean War in June 1950 – formalized the identity of Western Europe as a distinctive political entity. The externalization of political power gave rise to the need for integration while the Iron Curtain provided its immediate focus.

The arguments for new forms of European association were as much economic as political. Germany was clearly the key economy in Europe and had to provide the cornerstone of European reconstruction.

> The question was how to rehabilitate Germany so that she could make her economic contribution to the recovery of Europe as a whole, and yet prevent her from regaining her strength so far that she could once again be capable of posing a military threat to her neighbours. (Mowat 1973, 74)

The first step was simple: the policy of dismantling German factories had to be abandoned if Germany was to provide the dynamism for European economic recovery. The USA, at an early stage, realised that such a policy was obsolete. However, politicians in the UK and France were reluctant to act in the face of domestic public opinion, so that dismantling was only officially abandoned in 1949. That was only the first step in the reintegration of Germany: more important issues were still outstanding, including the future of coal and steel production which involved contentious elements such as the future of the Saar coalfield. This would become a critical factor in the movement to European integration.

A broader case could be made for integration in terms of Western Europe's position in the world economy. As recently as 1913 Western Europe had accounted for approximately one-half of the world's industrial production, yet, by 1950, this had declined to about one-quarter. It was apparent that it had fallen, and might continue to fall behind the two superpowers, the USA and the USSR. Regional economic integration provided *one* strategy for improving Western Europe's global competitiveness. The more general economic arguments for integration have been succinctly presented by Dunning and Robson:

> To overcome structural market distortions, such as tariff barriers, and to encourage competition.
> To reduce imperfections in foreign exchange, capital and labour markets.
> To facilitate product and process specialization by firms.
> To facilitate policy co-ordination in the circumstances of structural and policy interdependence.
> To develop economic and strategic strength by adopting common policies towards non-members.
> To increase market size and improve the technological capacity of member states. (1987, 105)

The same message was being preached, although in a more political form, by Jean Monnet as early as 1945:

> The nations of Europe are too circumscribed to give their people the prosperity made possible, and hence necessary, by modern conditions. They will need larger markets . . . Prosperity and vital social progress will remain elusive until the nations of Europe form a federation or a 'European entity' which will force them into a single economic unit. (Commission of the European Communities 1988a, 41)

Economic integration, however, can take many different forms. One specific form is a customs union, and it was this which was eventually adopted by the European Community. The advantages of such a union can be divided into static and dynamic effects (see Lintner 1989b). The static effects are the once-and-for-all benefits resulting from the creation of the union. They include *trade creation* resulting from increased internal free trade which, in turn, leads to economies of scale and lower prices. There is also *trade diversion*, being the negative effect of diverting trade from non-members to members of the union. In addition, there are the more uncertain but potentially more important dynamic benefits of a customs union. These include internal and external economies of scale of production, increased competition, and greater investment and technological dynamism resulting from increased growth. All these advantages were seen as being potentially available to an association of European states.

By the 1950s there were significant economic and political pressures for European integration. For some, the political arguments were pre-eminent: for example, the President of the EC Commission was to say in 1961 (EEC Press Release 22 May 1961) that 'we are not in business at all – we are in politics'. In contrast, Holland has said of the EC that:

> its originators drew on a prevailing ideology which was both economic and political . . . In essence, this was the ideology of liberal capitalism, or the assumption that the self-interest of enterprise could be harnessed in the public interest through a liberalisation of trade, capital and labour markets. (1980, 4)

It was the conjunction of these interlinked economic and political forces which was to make some form of European integration inevitable. The nation state had originated in Europe in the eighteenth and nineteenth centuries as a response to the needs of industrial capitalism. By the 1950s, it was clearly an inadequate political construction given the logic of capitalist accumulation and of geopolitics. Europe had arrived at a critical juncture.

The Route to the Treaty of Rome

While the necessary conditions for European integration existed, there were no sufficient conditions which dictated that the inevitable outcome would be the formation of the European Community. This can be seen in the events which led up to the signing of the Treaty of Rome:

1947   Benelux Customs Union established (Netherlands, Belgium and Luxembourg).
        Economic Commission for Europe established by the United Nations.
1948   The Brussels Treaty Organization formed as a mutual defence association by the UK, France, Netherlands, Luxembourg and Belgium. The Organization for European Economic Cooperation formed to distribute Marshall Aid.
1949   The North Atlantic Treaty was signed, initially by ten Western European countries in addition to the USA and Canada.
        The Council of Europe was set up 'to achieve a greater unity between its members for the purpose of safeguarding and realising the ideals and principles which are their common heritage and facilitating their economic and social progress'.
1951   The Treaty to set up the European Coal and Steel Community (ECSC), based on the Monnet and Schuman Plan, was signed by Belgium, France, FR Germany, Italy, Luxembourg and the Netherlands, and it encompassed the contested border region of Alsace-Lorraine. This was firmly backed by the USA as a way to expand the Ruhr's output of coal and steel at a time of rising global tension resulting from the Korean War.
1952   The Treaty to establish the European Defence Community was signed by the six original members of the ECSC, but was never ratified by France and did not come into force.
1953   A proposal for a European Political Community was rejected by France.
        The Beyen Report, which demonstrated the potential of a common market, was presented to and favourably received by the Foreign Ministers of the ECSC group.
1954   The Treaty to establish a Western European Union was signed as an extension of the Brussels Treaty Organization. It

had two new members, Italy and FR Germany, and facilitated the acceptable rearmament of the latter.

1955   The Messina Conference debated the future of European integration. The UK was poorly represented at this meeting which was to lay the foundations for the European Community.

1957   The Treaties of Rome were signed on 25 March establishing the European Economic Community and EURATOM, and came into operation on 1 January 1958. The membership was the same six countries that had come together to form the ECSC.

There was no inevitable causality in the sequence of events set out above. Instead, two different goals were evident in most integration-ist thinking during the 1950s: a military or defence union, and an economic union. The first critical moment in deciding which of these goals would be paramount came in 1951 with the creation of the European Coal and Steel Community (ECSC). It was the brainchild of the idealist Monnet and was carried through the French cabinet by Schuman, the Foreign Minister. Its genius was in offering advantages to both the main continental European powers, France and FR Germany. The German Chancellor, Konrad Adenauer, saw this as a political and economic turning-point for his country, offering rehabilitation, economic expansion and a measure of independence from the control of the superpowers. For France it offered a measure of joint control over resources in the face of any renewal of German aggression.

The establishment of the ECSC was important for the future creation of the EC in two ways. Firstly, despite initial difficulties, it was operational and demonstrated that economic integration was feasible. In no small part this was due to the highly pragmatic approach of Monnet and Schuman, who insisted on small-scale organizational steps which would yield concrete results rather than sweeping but impractical idealist schemes. It therefore served to point the way forward for further economic integration. Secondly, a Council, an Assembly and a Court of Justice were established for decision-making in and administration of the ECSC. This would eventually provide the model for the institutions of the EC and, in the meanwhile, would demonstrate that intergovernmentalism could work in Western Europe.

A second critical point was reached in 1954 when the French parliament refused to ratify the proposal for establishing a European army, and thereby killed off the scheme for a European Defence Community. With the proposal for a European Political Community also having been vetoed by the French parliament in the previous year, initiatives to organize European defence independently of the USA had largely failed. The Western European Union was established in 1954 but it was a weak organization and, instead, European defence became inextricably linked to the American-dominated NATO. Defence therefore ceased to compete as a serious basis for European integration.

By the time of the Messina conference, economic integration seemed the only realistic way forward for European federalists. The interests of the signatories of the Treaty of Rome were diverse (Commission of the European Community 1986b) but they all pointed to acceptance of economic union. For Germany there were opportunities for further political rehabilitation, the attraction of increased trade and the possibility of becoming the main centre of capital accumulation in the new Community. France also recognized the need for political reconciliation, as well as the potential for agricultural exports and for stimulating industrial growth. It was also attracted by the potential for using EURATOM as a way of sharing nuclear energy research costs in its bid to close the UK's European lead in this field. The Netherlands also stood to benefit from agricultural exports and hoped to secure a boost for its relatively small industrial base. Belgium, as an industrial exporter, saw clear trade advantages in a customs union, as did the small and already open Luxembourg economy. Finally, Italy perceived unique opportunities for export-led growth and for obtaining special assistance for the Mezzogiorno.

The notable absentee from the signatories of the Treaty of Rome was the UK. In the 1950s it saw itself as a centre of three interlinking circles: the Commonwealth, free Europe as a whole, and the Atlantic community. It had opposed most of the integrationist proposals of the early 1950s, such as the European Defence Community and the European Political Community, and was clearly ill-disposed to sharing power with any of its European neighbours. The Suez crisis gave a clear warning that the UK could no longer aspire to be an independent world power, but in 1956 and 1957 it had not yet shed such such aspirations.

Lack of democratic structures excluded some European states, such as Spain and Portugal, from EC membership. Others, such as Sweden, Finland, Austria and Switzerland, considered that their status as international neutrals was incompatible with EC membership. As a result, a second economic association was formed to represent the interests of some of these states (see Blacksell 1981). The Stockholm Convention of 1960 established the European Free Trade Association (EFTA) as a zone of industrial free trade: the founding members were the UK, Austria, Denmark, Norway, Portugal, Sweden and Switzerland. It was far less ambitious in scope than the EC, and did not encompass a common external tariff or common policies. It did share many of the European Community's goals of liberalizing capital accumulation, but without any significant loss of national sovereignty or commitment to sharing powers with the institutions of the new association.

It was probably true to say that, by 1956/8, events had led the original six members of the EC to the position where there was no obvious alternative future for association other than through their economic integration in the European Community. Once that position had been reached, a number of contingent events ensured that there would be rapid agreement on the Treaty of Rome. At one level, the Soviet crushing of the 1956 uprising in Hungary sharply underlined the need for greater Western European co-operation. But the most significant developments were in French politics, particularly as France was likely to be the most reluctant, yet the critical member of the EC. Two events finally led to French agreement to the Treaty of Rome. The Suez debacle demonstrated to France, as the UK, that its days as an independent superpower were numbered. Even more importantly, events were beginning to turn sour for France and the French government in Algeria. Meanwhile, de Gaulle was in the wings waiting to take power, and at that time he was known to be strongly opposed to European union. In such circumstances, the French government cast aside hesitancy and signed the Treaty of Rome, as one of the six founder members.

The Treaty of Rome: a Framework for the Future of Europe

The Treaty of Rome was mainly concerned with the economic framework of integration, establishing a customs union, and common markets for capital and labour. The signatories were also committed to development of common policies in a number of fields, including agriculture, transport and the European Investment Bank. However, it was inevitable that it would have to address both political and social issues. The Treaty was based on general principles and bold intentions, and no attempt was made to delineate policies in detail. This was unavoidable given the speed of the proposed integration programme and the enormous problems involved in reconciling the diverse interests of the member states. Not surprisingly, the founders chose a pragmatic step-by-step approach, and the first step was simply signing the Treaty. Even so, this committed the Six to a number of general objectives.

> *Customs union.* All tariff and other barriers to trade amongst the member states were to be eliminated. The earlier creation of the Benelux Customs Union meant that three of the members were already well advanced on such a programme. This was expected to generate trade-creation effects within the Community.
> *Common external tariff* to be formulated between the Six and all other states and international trade groups. This was expected to lead to trade diversion effects to the benefit of the member states.

Free market for labour. Articles 48–58 guaranteed the freedom of movement of workers, the equal treatment of all workers with respect to wages and conditions, entitlements to welfare benefits, and the right of establishment for the self-employed. *Free market for capital* involving the ownership and movement of various forms of capital, rights of establishment, and transnational sales of financial services. *Common policies* to be established including a *Common Agricultural Policy*, a *Social Fund*, and a *Common Transport Policy*. *European Investment Bank* financed by the member states and empowered to provide loans for economic restructuring within the Community.

These were only the major provisions of the Treaty of Rome. However, the Treaty in detail had the same emphasis on negative integration measures as was evident in the major aims. The Treaty of Rome was mainly concerned with advancing the interests of liberal capitalism: as such, its major objective was the elimination of barriers to the operations of free markets, whether for capital, labour or products. It was firmly committed to deregulated and (internationally) unfettered, private capital accumulation as the major engine of economic integration. Consequently, as Holland (1980, 10) argues, the Treaty is 'mainly concerned with preventing abuses to competition and the market mechanism rather than with providing a framework for joint intervention to achieve what the market itself cannot do'. There were a number of reasons for this negativism but, essentially, it stemmed from the minimalist approach which has characterized most of the Community's activities. The creation of a customs union and the elimination of international trade barriers were relatively uncontentious, compared to the questions of social distribution and public sector policies. Given the speed with which the Community was established following the Messina Conference, it was highly unlikely that more detailed, more positive or more contentious policies could have been negotiated in such a short time span. This was especially so given the very different political philosophies prevailing in the member states,

Table 2.1   Western European trading blocs in 1958

	% manufacturing exports to	
	EC6	EFTA8
EFTA Group*		
Norway	25.5	39.1
Sweden	21.3	38.7
Denmark	24.5	37.9
Portugal	12.4	14.4
UK	12.2	9.8
Austria	36.5	13.2
Switzerland	37.2	16.3
EC Group*		
Belgium	42.3	15.1
Netherlands	39.1	22.8
France	19.5	12.4
Italy	20.8	16.3
FR Germany	23.4	29.1

* No data on Ireland or Luxembourg
Source: United Nations Commodity Trade Statistics; after
Wijkman (1989)

and particularly the West German opposition to many forms of state
interventionism.

The establishment of the European Community drew a new line
across the economic and political maps of Europe. Although not as
rigid or exclusive as the Iron Curtain, it was to reshape the whole
pattern of trade, investment and decision-making within Western
Europe. Both the EC and EFTA cut across existing patterns of trade
in Western Europe; they were creating new economic zones rather
than formalizing existing, discrete trade networks (see table 2.1).
Within the EC6 only Belgium and the Netherlands had significantly
stronger trade links with other members of this group than with
EFTA. FR Germany, which was the economic cornerstone of the
Community, actually had stronger trade links with EFTA. The

EFTA8 group had even less internal coherence and only the Scandinavian members were locked together by a strong trade network. In contrast, both Austria and Switzerland had stronger links with the EC6, not least because of their dependence on the German market. The UK also had stronger ties with the EC than with EFTA but, more importantly, had relatively weak trade relationships with both groups because of its Atlantic and Commonwealth alliances. Therefore, the trade diversion and creation effects of these two associations were likely to have major effects on the economic geography of Western Europe.

The creation of the EC was innovative in political as well as economic terms. The member states had to concede some of their national sovereignty but, at the same time, conditions clearly were not yet ready for federalism. The compromise was intergovernmentalism, but within the legal framework of the EC (Wallace 1982; Taylor 1983; Harrop 1989). Four major institutions were created: Parliament, Court, Commission and Council of Ministers. However, they did not possess equal powers in the system of checks and balances which they sought to establish.

- *The Commission* provided the bureaucracy of the EC. It proposed new policies, implemented existing policies, was the Guardian of the Treaty of Rome and had a formal right to attend Council meetings.
- *The Council of Ministers* took the major decisions with respect to the detailing of existing policies and principles and to the adoption of new policies.
- *The European Parliament* had a mainly consultative role but it did have the power to approve the Community budget.
- *The European Court of Justice* had the responsibility of assessing the actions of the other institutions and of the member states against the Treaty of Rome and other fundamental principles.

Over time the balance of power between the institutions was to change. However, ultimate power rested with the Council of Ministers, which was composed of the Heads of Government of the member states. The very act of signing the Treaty of Rome implied some loss of national sovereignty; for example, in determining external tariffs, or being subject to challenge by the European Court on breaches of Community law. But most major decisions still required the consent of all the Heads of State. As in the agreement on the terms of economic integration, gradualism and pragmatism were also the key to success in the political realm. 'Rather than relinquish all sovereignty overnight, the Member States were asked merely to abandon the dogma of its indivisibility' (Commission of the European Community 1986b, 22). However, such an essentially pragmatic approach created a hybrid political organization which, in a sense, was 'neither above or below the state' (Wallace 1982). The power given to the Council of Ministers institutionalized the privileged positions of the individual states. As a result, the key to understanding the policies and the subsequent evolution of the EC lies, above all, in an analysis of *domestic* politics. Ross (1991, 65) puts this in a broader context when he argues that 'Struggle over the future of Europe is largely about developing more promising environments for capitalist success, but it is being carried out by political actors in a relatively open-ended arena'.

THREE

Constructing the European Community: 1958–73

Introduction: a Unique Opportunity

The EC was established in the late 1950s in the midst of the most favourable period of economic growth in the twentieth century. Global and European expansion smoothed the way for the introduction of a new trading framework, of common markets in labour and capital, and of fledgling common policies. The exceptional growth context also eased the processes of political compromise and bargaining which were to become the hallmark of decision-making in the Community. Even so, domestic politics were to be critical in shaping common policies, especially for agriculture. In turn, the Common Agricultural Policy would influence profoundly the future evolution of the EC.

While remarkable progress was made in constructing the EC during this period, the dynamism of the Community should not be exaggerated. Barraclough (1980, 64) has argued that 'the initial successes were deceptive. Progress towards integration and towards what was optimistically called a "New Europe" went ahead so long as conditions were favourable.' The creation of the EC contributed to these conditions but it also benefited substantially from them. All of Western Europe enjoyed strong economic growth during the 1960s, as can be seen from table 3.1. If the sluggish UK economy, and the rapidly industrializing Southern European economies (see Williams 1984, introduction) are excluded from the picture, then there is little difference in the performances of the EC and of the non-EC Western European economies during this period. It remains an open question whether the EC sustained these high growth rates

34

Table 3.1 GDP per capita growth rates in Western Europe, 1960–70

EC	% annual growth rates	Non-EC (Northern)	% annual growth rates
Belgium	4.8	Austria	4.5
FR Germany	4.4	Denmark	4.7
France	5.7	Finland	4.6
Italy	5.3	Ireland	4.2
Luxembourg	not available	Norway	4.9
Netherlands	5.5	Sweden	4.4
		Switzerland	4.3
		UK	2.9
Non-EC (Southern)			
Greece	6.9		
Portugal	6.2		
Spain	7.1		

Source: World Development Report 1985, New York: World Bank

or was carried forward by them. In a sense, however, the last question had become irrelevant by the early 1970s for, by that stage, the EC had become an established economic structure. Major changes had occurred in trade and investment within Europe, thereby restructuring the economic geography of the region. The EC, in 1958, had been arbitrarily constructed as an economic zone but, by the late 1960s, it had become a highly significant feature of the Western European economy. Not least, the trade diversion and creation effects generated pressures on both member and non-member states which contributed to the first enlargement of the EC on 1 January 1973.

The economic success of the EC occurred despite a troublesome period in its construction in the mid-1960s. One of the major difficulties lay in forging the Common Agricultural Policy, the area where domestic political pressures were most intense. The inevitable political compromise resulted in the creation of a common policy which has dominated the budget and all other policy areas throughout the existence of the EC. Another and linked difficulty was the establishment of policy-making procedures. One of the founding principles was that there would be majority voting in almost all policy areas. However, the loss of sovereignty that this implied was to prove unacceptable, especially to France. This led to the 1966 Luxembourg compromise which established unanimous voting as the basis for decision-making. While this decision eased the task of shaping the EC in the 1960s, it was to prove a major obstacle to further economic and political integration in the 1970s and 1980s.

The Customs Union and the Common External Tariff: a New Framework for Trade

The customs union and the common external tariff did not come into existence immediately after the signing of the Treaty of Rome. Some tariffs were cut on 1 January 1959 but a long transitional period was necessary to negotiate and implement the full programme. Differences in national interests – especially between France and FR Germany – ensured that this was a troublesome process, but it was successfully completed in 1968, eighteen months ahead of schedule. In retrospect, it seems that the dismantling of tariff barriers was assisted by favourable external circumstances. The economic growth of the 1960s has already been referred to: this meant that if some of the industries of member states proved to be uncompetitive in the single market for products, there were favourable opportunities for national resources to be switched into other growth sectors. In other words, the unemployment implications of restructuring rarely had to be faced in the 1960s. Secondly, despite some divergence, the six founder member states had broadly similar competitive positions and trading interests in 1958. The exception was Southern Italy, but it was expected that this region would benefit from special assistance programmes. This communality of interest greatly assisted the process of political compromise. Finally, the broad consensus of economic interests was underpinned by a strong international convergence of tastes and culture during the consumer boom of the 1950s.

Of course, the creation of the customs union did not remove all barriers to trade between the Six. Most importantly, it only applied to (most) industrial goods. The Treaty of Rome (Article 59) had also specified that there should be progressive abolition of restrictions on the freedom to provide services within the Community, especially with respect to finance, insurance and banking. Yet 'for more than twenty five years after the EEC Treaty was signed, progress in financial integration was discontinuous, uneven, and on the whole modest' (Micossi 1988, 217). Furthermore, the liberalization of trade was restricted even with respect to industrial goods, for there were still non-tariff barriers to be overcome. Indeed, this is one of the central concerns of the 1992 programme. Nevertheless, an important start had been made in constructing a new Western European trade framework.

The importance of this is evident in the trade statistics for the period. As has already been seen (see table 2.1), the EC and EFTA were relatively arbitrary trade groupings when initially established. Not least, this was because the UK had stronger trade links with the EC than with EFTA, while FR Germany had a pivotal position in the trade of both groups (Wijkman 1989). As a result, the customs union and the common external tariff had a profound effect on the pattern of trade. For the member states, trade was much more dynamic within the EC than with non-members (Dunning and Robson 1987, 111). This is illustrated by the import statistics for 1958 and 1972 (see table 3.2). The overall dynamism of the Community is shown by the way its share of total world imports increased from 21.1 per cent to 28.6 per cent during this period.

Table 3.2   EC6 imports, 1958–72

		1958	1972
A	Imports as % world trade	21.1	28.6
B	Intra-imports as % of all imports	29.6	51.6

Source: Reichenbach (1980)

However, while the global share of intra-EC imports almost doubled to 19.4 per cent, the share of extra-EC imports fell sharply. As a result, intra-EC imports as a share of *all* EC imports increased dramatically from 29.6 per cent to 51.6 per cent. Not only was the process of capital accumulation dynamic and increasingly internationalized, but it contributed significantly to the integration of the economies of the Six.

Even in the absence of the EC, global economic growth would probably have led to increased dynamism in Western Europe and to greater integration amongst some or all of it states. However, the creation of a customs union and a common external tariff did have trade creation and trade diversion effects (p. 24) for the Community. There are major difficulties involved in quantifying these effects. However, Harrop's (1989, 51) review of the econometric evidence suggests that the best estimate is of trade creation effects of $10 billion and trade diversion effects of $1 billion. Even though trade diversion was the lesser of the two effects, it generated conflict with other major trading groups. Special agreements with EFTA cushioned other Western European states to some extent, so that the USA considered itself to be the most aggrieved party. As early as 1963 this led to a major trade conflict. During the so-called 'chicken war' the USA imposed levies on EC exports of brandy, VW trucks and potato starch in retaliation against EC import levies on American poultry.

Capital, Labour and Internationalization

The EC6 was an association of some of the wealthiest economies in Western Europe. It was also relatively homogeneous, especially in contrast to the new member states added by the 1970s enlargement (see table 3.3). If the mean GDP per capita (at purchasing power parity) for the EC9 group in 1960 is taken as the standard for comparison, then only Italy of the original Six fell significantly below this level. FR Germany had the highest level of GDP, followed by the Netherlands and France. The first two decades of

Table 3.3 GDP per capita at purchasing power parity in the EC9, 1960–79

	EC9 average = 100	
EC6	1960	1979
FR Germany	118	118
Netherlands	104	103
France	100	112
Belgium	98	108
Italy	69	71
Luxembourg	NA	111
EC9 enlargement		
Denmark	113	116
UK	112	91
Ireland	59	61

Source: Hallett (1981, 26)

growth within the EC framework did little to change this overall picture. In 1979, FR Germany still had the highest level of GDP per capita, closely followed by the other members, with the exception of Italy which continued to lag more than 30 per cent below the average. At the same time, the poor performance of the UK is reflected in a substantial fall in GDP relative to the EC average.

These data, however, do not convey the true extent of German and, to a lesser extent, French economic dominance within the EC. The West German economy expanded at high relative and absolute rates throughout the 1960s and 1970s, and it eventually became the largest industrial exporter in the world. There were already signs of this remarkable trading performance during the first period of the EC's existence. By 1973, 28.5 per cent and 18.0 per cent, respectively, of the intra-EC exports of the Community originated from Germany and France. Similarly, almost one-half of intra-EC imports were accounted for by these two countries (Ziebura 1982, 132). These data, much more than those for GDP per capita, indicate the considerable weight of the two economies within the original Six. They also point to one of the fundamental reasons why France and Germany became *the* decisive influences on the way that the Community evolved during this period. Germany, in particular, had become the motor of economic growth within the EC. Later, as the UK economy continued its long-term relative decline, FR Germany was to take over the economic leadership of Western Europe as a whole.

The changes in national economic power were underlain by major processes of capital restructuring. The post-war years witnessed a remarkable internationalization of capital in response to the escalating costs of research and development, diversification strategies, the need to secure market access and the reorganization of the labour process (see Williams 1987, 68–71). As a result, the global stock of foreign direct investment (FDI) increased from $108 billion to $287 billion between 1967 and 1976 alone (Sauvray 1984). The main driving force behind this internationalization was the expansion of American transnational corporations. American hegemony in the first two post-war decades is reflected in the increasing share of global FDI which originated in the USA: only 35 per cent in 1930 compared to 52 per cent in 1971. There was a corresponding decline in the proportion of FDI which originated in Western Europe, especially as a consequence of the disruption caused by the Second World War. The UK share fell from 44 per cent to 15 per cent, while that of France, the EC leader in FDI, fell from 8 per cent to 6 per cent.

As Western Europe lost its position as the major global exporter of capital, it became the principal recipient of American FDI. At first, the UK was the major point of attraction. This was hardly surprising given that English was the language of international business, the financial importance of the City of London, and Britain's role as the centre of European and Commonwealth trade networks. However, the creation of the European Community in 1958 led to a significant increase of investment into the member states (Darby 1986). Whereas, in 1950, the UK received one-half of all US investment in Europe, by 1975 this share had fallen to 28 per cent. Although the UK was still the leading destination, FR Germany was closing the gap and received 18 per cent of USA FDI in Europe by the latter date.

American investment in Europe involved acquisitions and greenfield site development. Both of these caused European concern about what Servan-Schreiber (1968) labelled 'the American challenge'. The greater size, vision and technological lead of American transnational companies enabled them to develop more genuinely multinational strategies within Europe than the Europeans achieved. French, German and British companies still focused most of their operations on one or two countries. Consequently, it was American rather than EC transnational companies which benefited most from the creation of the Community.

Yet one of the major aims of the customs union and the common external tariff had been to secure long-term dynamic benefits for EC companies. Larger markets were supposed to lead to economies of scale, greater efficiency, and improved research and development (R&D) capacity. The latter was seen as particularly important as American companies were known to spend four times as much as EC companies on R&D. Walter Hallstein of the EC Commission stated that 'scientific research has been called the third factor affecting production in the modern economy after capital and labour' (1972, 198–9). In this context, the ability of American multinationals to enjoy the advantages of the customs union, and to establish trans-EC sales and production networks, must have appeared highly ironic to the founders of the Community.

Did the increased American presence imply that the EC's industrial strategy had failed? In a sense the answer is no, for an increase in FDI was the inevitable outcome of the formation of the common external tariff. Non-EC firms located production within the Community to overcome the barrier presented by the external tariff. Furthermore, many of the direct benefits of the customs union had been realized, especially in terms of the scale of European companies. For example, the fifty largest European companies accounted for 25 per cent of Europe's gross industrial output in 1976 compared to only 15 per cent in 1965 (Geroski and Jacquemin 1985, 172). But scale economies did not automatically translate into the hoped-for dynamic benefits of the customs union. In part, this was because European companies did not develop genuinely trans-EC strategies. Instead of scale economies leading to specialization (and hoped-for research and technology gains), the main effect was harmonization: industry was developed along similar lines by national capital within each member state. This duplication of industrial structures was to contribute to the severity of the 1980s global crisis in Western Europe, and eventually would provide a major stimulus for the 1992 programme.

There were also limitations resulting from the arbitrary nature of the EC6 as an economic zone, in the face of the increased globalization of capital. Formation of the EC led to a spate of EC-wide international mergers, with 257 being recorded between 1961 and 1969. However, these were dwarfed by the 563 mergers between EC and non-EC companies (Holland 1980, 68). Here lies the key to interpreting the industrial restructuring of the 1960s. American companies had taken more immediate advantage of the creation of the EC but, in a sense, this was one of the final expressions of US hegemony. It was also true that EC companies had not international-ized as extensively within the Community as had been hoped for. However, EC companies had been heavily involved in global strategies of internationalization. Their overall growth rates were also slightly higher than those of American companies during the 1960s (Rowthorn and Hymer 1970). The stage was being set – partly assisted by the creation of the EC – for Western European companies to challenge American hegemony of the Atlantic, if not the global, economy during the last quarter of the twentieth century.

It was not only capital which was being restructured during this period; there were also major changes in labour markets. In particular, there was a lack of labour supply constraints, which Kindleberger (1967) considered to be the key to the virtuous circle of growth in Western Europe in the 1950s and 1960s. There were several sources of labour supply. National sources included increased female participation rates, and the release of labour from the restructuring of agriculture (as in Western France) and of traditional mining and manufacturing (as in the Ruhr). The process of restructuring was linked to the generation of inter-regional labour migration flows to the national centres of rapid capital accumulation.

International migration was another source of labour supply. This was important to the process of capital accumulation in the more developed economies of Western Europe in a number of ways: the elimination of supply bottlenecks, the realization of scale economies, reduced costs, and increased exports resulting from reductions in the prices of internationally traded products (Paine 1977). One of the aims of the EC had been to maximize such benefits by creating a common market for labour. Hence, the Treaty of Rome stipulated not only that there was to be freedom for EC citizens to take jobs anywhere in the EC, but also that there was to be no discrimination against non-nationals with respect to pay and conditions of work. These rights were gradually established during the first decade of the Community's existence, culminating in a key Council Regulation in January 1970 (Molle and Van Mourik 1988).

In practice, the creation of a common market was to prove less important for labour migration within the Community than had probably been expected. The number of intra-EC6 international migrants did increase from 576,000 in the late 1950s to 905,000 in 1974 (Straubhaar 1988). Approximately two-thirds of the total were from Italy at both dates. However, intra-EC migrants were substantially outnumbered by migrants from outside the Community on both occasions (see figure 3.1). Indeed, by 1974 there were approximately 3,500,000 extra-EC migrants compared to only 905,000 intra-EC migrants; the latter represented only 22 per cent of the total. The rapid process of capital accumulation in the Community during these years could only be sustained by drawing on the vast labour reserves of the Mediterranean region as a whole, not just of Italy. For example, by 1968 Greece, Spain and Portugal provided 29 per cent of international migrant workers in the EC6 compared to only 13 per cent in 1958 (Molle and Van Mourik 1988, 323). *Gastarbeiters*, or guestworkers, were also drawn from Turkey, Algeria and other North African states, as well as from the ex-colonies of France, Belgium and the Netherlands. Figure 3.2 shows the growing importance of international migration in all the EC6 economies during the 1960s. Excepting Luxembourg, dependence was greatest in France and FR Germany, with 11 per cent of their

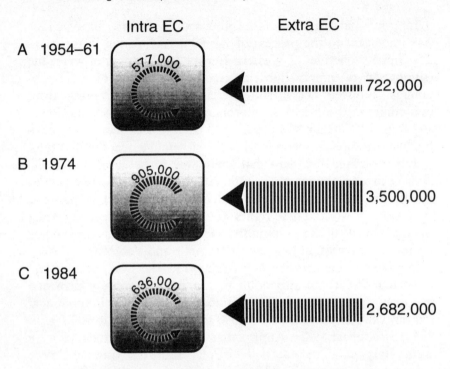

Figure 3.1   Sources of emigrants in the EC6, 1954/61–84
Source: Straubhaar (1988, 59–61)

workforces being immigrants in 1974. International migration –
even if not from the EC – was clearly crucial in the growth of these
two economies.

The relationship between international trade and migration is
complex. Straubhaar (1988, 48) argues that there were three
posssible outcomes of the formation of a common market. The first
of these was a decrease in intra-EC migration and an increase in
intra-EC trade. This was certainly the optimistic, hoped-for long-
term outcome of the diffusion and strengthening of growth
throughout the Community. Secondly – and clearly not the expected
outcome – an increase in intra-EC migration and a decrease in intra-
EC trade, as labour migration was substituted for trade. Finally, it

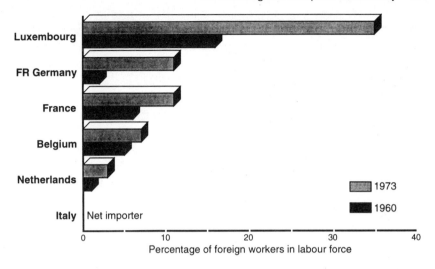

Figure 3.2   EC6 net dependency on foreign labour, 1960–73
Source: Molle and van Mourik (1988, 324)

Table 3.4   Migration and trade flows in the EC6 and EC9,
1958–80

| | Percentage annual worth | |
	1958–1973 (EC6)	1973–1980 (EC9)
Migration		
Intra-EC	2.4	− 12.5
Extra-EC	13.6	− 14.1
Trade		
Intra-EC	18.6	22.7
Extra-EC	12.4	24.7

Source: Straubhaar (1988, 51–3)

was possible that there would be an increase in both migration and
trade within the EC, which was probably the realistic expectation for
the short term. Straubhaar examined this relationship in terms of
relative intra- and extra-EC flows rather than of absolute levels (table
3.4). In the first period, 1958–73, intra-migration grew less rapidly

than extra-migration and intra-trade grew more rapidly than extra-trade. Consequently, he concludes that the first of his hypotheses was correct: trade and migration were in a substitutional relationship, at least in relative terms.

In more general terms Straubhaar's work confirms that the liberalization of labour migration laws did not significantly stimulate intra-EC migration compared to inmigration from outside the common labour market, even in the short term. As Holland argues: 'it was not the logic of economic integration as such but accumulation of capital and its demand for labour which attracted migrants from less to more developed countries' (1980, 58). In other words, the restructuring of labour markets in the EC, as of capital markets, was only a part of the larger globalization of the European economy.

Common Policies and Political Compromises

The European Community was conceived of as more than a customs union and a common market for capital and labour, for there was also a commitment to developing common policies. As these were not specified in detail in the Treaty of Rome, they were to be worked out in practice during the early years of the Community's existence. As the common policies would have to be forged in the context of intergovernmentalism and entrenched national interests, it was inevitable that the process would be characterized by political compromise. This was clearly evident in relation to the Common Agricultural Policy (CAP), the most politicized as well as the most important of the common policies.

When the Treaty of Rome was signed, agriculture was still a major industry in most European states. For example, it accounted for 35 per cent of all employment in Italy, and 25 per cent in France. However, nothing underlines the importance of the CAP as much as the significance that de Gaulle attached to it. Soon after his election to the French presidency in 1958, he called an executive meeting to decide whether France should remain a member of the EC. The critical argument, accepted by de Gaulle, was that the potential benefits of the CAP would compensate for any industrial trade losses, especially with Germany (Holland 1980, 27).

The detailed CAP negotiations were fraught with difficulties. The Treaty of Rome had set down the aims of the CAP in Articles 34–87: to increase productivity, to provide a fair standard of living for the agricultural community, to stabilize markets, to assure the availability of supplies and to supply consumers at reasonable prices (see Hill 1984). These aims were contradictory and there were also numerous ways in which they could be implemented. However, the three main features of the policy framework were quickly identified: free trade within the EC and a common customs barrier against the rest of the world, guaranteed prices and structural reforms. But this still left considerable detail to be finalized.

Probably the major obstacle was the very different national farm policies which had evolved in the member states since the nineteenth century. For example, France, Germany and Italy had highly protectionist policies, while the Netherlands and Belgium had liberalized trade which had resulted in increased specialization and competitiveness. The inter-war depression had led to greater protectionism throughout Europe, but there were still major differences in the agricultural interests of the member states in the 1950s. The Netherlands wished to expand a modern, exporting farm sector; Belgium wanted defensive farm subsidies; while Germany, France and Italy sought to protect their marginal, fragmented farms (Blacksell 1981). The negotiations were further complicated by the diverse interests within each of these countries; for example, the French government had to balance the requirements of large-scale farming in the Paris basin against those of a still predominantly peasant agriculture in the South and the West.

Negotiations quickly became embittered, especially between the two leading members of the Community. By 1961 de Gaulle was threatening that France would not agree to proposed cuts in industrial tariffs until the framework for the CAP had been established. Not for the last time, the CAP seemed to threaten the very existence of the EC. By January 1962 the EC had to engage in 'clock-stopping' to maintain the deception that it was still keeping to agreed timetables for the CAP negotiations. However, the general principles of a price system were eventually agreed in January. The guarantee system established target prices for farm products, and if these were not achieved in the market, then the EC would buy up surpluses at intervention prices. There were also thresholds to bring the prices of imports up to the guarantee levels, and export restitutions to subsidize exports. The January 1962 agreement covered about 40 per cent of the produce of the EC including dairy and pigmeat products. However, in order to secure German acceptance of the CAP, prices had been set close to those prevailing in its national system, and these were much higher than world market prices. The CAP therefore came into existence as a high-cost policy, and this has been a constant feature subsequently.

The CAP saga was still not over, for individual prices and intervention mechanisms had to be agreed for each separate farm product. By 1965 the Community was again in political crisis as France refused to accept the principle of majority voting by the member states, especially in relation to the CAP. This led to the 'empty-chair policy', as France refused to sit at the negotiating table until unanimous voting was accepted in the 1966 Luxembourg compromise. Thereafter, fairly rapid progress was made in finalizing the CAP. The cereal price regime came into existence in 1967 but with prices set much higher than both the EC average and current world levels. This reflected, above all, the interests of German farmers, although they still had to accept 15 per cent price cuts (Hendriks 1989). Common agricultural markets were then established in rapid succession for milk, sugar beet, fruit and vegetables. However, the political compromise which had established such high prices for cereals had set an important precedent. High prices would also have to be set for all other farm products, if only to maintain price differentials. The last major price agreement – covering wine – was reached in 1970 so that, in total, it took almost a decade to flesh out the CAP.

The CAP has been a mixed economic success. It certainly boosted output, which is not surprising, given that target prices were set well above world prices throughout the late 1960s and most of the 1970s and early 1980s. As a result, self-sufficiency in the EC rose markedly; for example, from 90 per cent to 114 per cent for wheat between the late 1950s and 1980 (Hill 1984). The guarantee policies have also helped to stabilize prices and to provide reasonable standards of living for farmers. However, the cost has been borne by the consumer in terms of higher prices. The EC's general policy-making capacity has also suffered from the large and increasing proportion of the EC's budget which has had to be devoted to financing the CAP.

It is difficult to decide who benefited most and who least from the creation of the CAP. It was certainly true that France made least and FR Germany made most of the necessary political compromises. Perhaps the real winner was the Commission; its central role in forging the compromise reinforced its standing and authority amongst the other institutions. At the same time, the Luxembourg compromise, which was an integral part of the CAP political bargain, was to have profound – and largely negative – effects on the future decision-making capacity of the Community. The most damning indictment comes from Holland (1980) who argues that 'the establishment of the practice of unanimous voting for several years translated the Community from a venture in supranationalism into an inner OECD with the incubus of the common agricultural policy'.

The other areas of common policy proved less contentious, if only because the budgetary implications of the CAP left little scope for resource redistribution elsewhere.

• *The Social Fund* was potentially one of the more important common policies. The Treaty of Rome saw this as an instrument for easing some of the short-term labour market difficulties which could result from the creation of common markets and the customs union. Hallstein argued that: 'The provisions of the Treaty of Rome on social policy – like those on transport policy, monetary policy and other matters – clearly bear the mark of the controversies during the negotiations leading to the conclusion

of the Treaty. In other words, they bear the mark of the past'
(1972, 169). The desire of each country to protect its own
distinctive social policies meant that few concessions or agreements were possible either during the negotiations, or subsequently. As a result, there was little development of EC social
policies in the 1960s. In practice, most of the Social Fund
expenditure was directed at retraining measures to ameliorate
some of the consequences of industrial restructuring. The total
spending even on this was limited to about £120 million by 1970.
Such a policy could only have a marginal impact on the social
needs arising, with increasing frequency, from labour market
adjustments towards the end of this first period. Even less
progess was made with other common policies.

- *The European Coal and Steel Community* was heavily involved in
  the restructuring of the coal industry, and EURATOM co-
  ordinated jointly funded research on nuclear-generated electri-
  city. However, there was no coherent policy for the energy sector
  as a whole during this period and no serious attempt was made to
  develop one until after the first oil crisis (Odell 1976). Energy,
  like social policy, was a jealously guarded area of national
  sovereignty.
- Progress was also slow with respect to the *Common Fisheries
  Policy*, even though the Treaty of Rome specified that a common
  market should be established for fish products. Again, this was
  an area of strongly competing and diverse systems of national
  protection. Eventually agreement was reached in 1970 on the
  common organization of markets, a structural (modernization)
  policy, and qualified open access to all EC fisheries. However,
  this was only signed in the shadow of negotiations over EC
  enlargement. This contributed to the mutual self-interest of the
  Six in establishing the Common Fisheries Policy as part of the
  acquis communautaire (general policy framework) which would
  have to be accepted as non-negotiable by any new members.
- *The Common Transport Policy* was even more neglected and it
  remained in what Hallstein (1972, 226) labelled 'pastoral

seclusion'. There were no attempts to liberalize or co-ordinate air and sea transport and only marginal intervention in land transport. Such issues would not seriously be considered until the 1980s.

- There were also many neglected areas of economic policy, including *monetary policy*. In the late 1960s the EC was rocked by three major currency crises: sterling in 1967, the dollar and the French franc in 1968. The last of these, in particular, was a source of serious potential disruption to policies and trade within the EC. Again the relationship between the two dominant economies was crucial: FR Germany refused to follow the French devaluation and instead preferred to cut import taxes by 4 per cent. In effect, there was no real co-ordination of national monetary policies. The crisis led to the Council of Ministers' agreeing the Barre Plan in July 1969, which proposed to increase monetary union and economic policy co-ordination. This initiated a series of attempts to construct monetary union during the following two decades.

- *Competition and mergers* were another difficult area of economic policy. The EC had contradictory aims in this area. On the one hand the Treaty of Rome emphasized that competition was paramount, being seen as a motor of improved efficiency and competitiveness. On the other hand, it wished to promote the creation of European super-companies, capable of competing against the strong and growing American presence both in Europe and in world markets. The 1970 Colonna Plan did, in fact, outline a possible policy framework for mergers and acquisitions, but was never adopted. Instead, the Council concentrated on eliminating anti-competitive practices. This is a difficult area to police, especially at the international level. Where persuasion failed, the Commission turned to the Courts. For example, in one of its classic early legal judgements, the European Court of Justice ruled that the EC Commission had been right to declare as void an agreement between Grundig and Costen to limit Grundig's sales in France.

- A *common regional policy* was not established, although there were considerable regional inequalities in the EC, and it seemed that integration had contributed to these.

As the EC common policy framework evolved, it quickly became apparent that it was likely to be heavily circumscribed by budgetary constraints. The EC's own resources were derived from a fixed percentage of the VAT returns of national governments (subject to an upper limit). As the Community came into operation, so its expenditure increased sharply: from 35.5 million ECU in 1958 to 607 million in 1965, and then a further fivefold increase to 3331 million ECU in 1972 (table 3.5). There were also major shifts in the allocation of resources. In 1965, the CAP was still being negotiated and it only accounted for 16.9 per cent of the total EC budget, but by 1972 it consumed three-quarters of total Community expenditure. The early process of political compromise had produced an expensive farm policy which was to influence the future evolution of all other Community common policies. With the CAP, the EC had created one substantial and genuinely Community-wide policy which was fully operational by the end of the 1960s. But the price for this was high: the severe limitation of the possibilities for many other

Table 3.5   The EC budget, 1958, 1965, 1972

% expenditure on:	1958	1965	1972
European Development Fund		41.0	6.4
EURATOM	22.3	19.8	
EAGGF (CAP)		16.9	74.4
Social Fund		7.1	2.9
Industry, Energy, Research			2.3
Administration	16.6	9.1	12.7
ECSC	61.1	6.2	1.3
Total	100.0	100.00	100.0
Absolute expenditure (million ECUs)	35.5	607.2	3,330.9

Source: Commission of the European Communities, *European Economy*, Annual Economic Review, 1986–1987, no. 29, July 1986, 166

initiatives such as a positive industrial strategy, or a comprehensive social policy. The longer term political costs were to prove even greater, for the costs of the CAP, and its general budgetary consequences, were to shackle decision-making throughout the following years.

The First Enlargement: the Crowning Achievement of the First Period?

The first fifteen years of the Community's existence had been a difficult period. Many areas of decision-making had been contentious and had, as with the French 'empty-chair' strategy, stretched the credibility of the EC. However, this should not be allowed to detract from the achievements of these years. In 1957 the Treaty of Rome was signed as a general statement of intent rather than as a detailed and polished policy document. The Six had little more to draw on than political goodwill and their shared experience of the ECSC. They aimed to establish a customs union, a common external tariff, common markets for capital and labour and a corpus of common policies, within a relatively arbitrarily-defined economic zone. Yet, within fifteen years substantial progress had been made in all these areas and the Community of Six was to be enlarged to a Community of Nine. The accession of the UK, Eire and Denmark was to make the EC the central economic grouping in Western Europe. It encompassed virtually the whole of North-west Europe, and only the Mediterranean, Scandinavian and Alpine regions remained outside its boundaries.

The EC/EFTA split was always an artificial division within Western Europe, especially because of the pivotal-trade links of the UK and FR Germany. Consequently, there was strong pressure to restructure the relationships between the two groups from the very beginning. The success of the EC in establishing the bases for common policies, and the relative strengthening of the German and French economies compared to the UK, served to intensify these pressures. The continuing internationalization of capital accumulation was also pointing to the need for further political reorganization of the European economic space. Constitutionally, any such enlargement would be quite simple, for Article 237 of the Treaty of Rome stated that 'Any European state may apply to become a member of the Community.'

While both Greece and Turkey sought and obtained special association arrangements with the EC, the critical relationship was with EFTA. The UK, which had been the driving force behind the creation of EFTA, soon sought to transfer its allegiance to the EC. Together with Denmark, Ireland and Norway, it applied for full membership of the EC but was promptly rejected in 1963, largely at the insistence of the French President, de Gaulle. While it was true that more time was needed for the evolution of the fledgling institutions and policies of the EC, there were other reasons for this rejection. Not least, UK membership might challenge the central role of France and Germany in shaping the Community in its critical early stages. In addition, the strong ties of the UK with the USA might compromise the attempt to establish the EC as an independent economic and political force.

In 1967 the four prospective applicants reapplied for membership. The UK was again the prime mover. Its economic motives were clear – GDP per capita growth rates in the EC in the period 1950–65 had been double those of the UK. In addition, the EC – especially FR Germany – was beginning to challenge the UK as the principal destination of European inward investment. The British market, on its own, and the larger EFTA market were too small to combine the scale requirements of R&D with the need for competition in sectors such as cars and consumer durables. The links between market size, technology and productivity were becoming increasingly transparent and there was a strong feeling that an independent UK – or EFTA – would not be able to compete at the appropriate level. The UK, belatedly, had come to similar conclusions as had the Six in the 1950s. There was also a strong element of wishful thinking that EC membership would provide a panacea for the increasingly obvious weaknesses of the UK economy.

Political considerations were also important in the UK's application. Decolonization was changing and weakening the role of the UK in the Commonwealth during the 1950s and 1960s. In addition, Suez was only one of several incidents which served to demonstrate that the UK could no longer aspire to the status of an independent super-power. It had little more than the role of an extra on the world stage in the global contest between the USSR and the USA. Therefore, in terms of its three historic circles of influence (see pp. 27–8), the only realistic alternative was the European option. It was equally clear that the EC might be able to provide this but that EFTA could not.

After lengthy negotiations the first enlargement was agreed in June 1971. The UK, Denmark, Ireland and Norway were to become members of the EC on 1 January 1973. After a referendum, Norway decided not to take up membership: there was too much unease about the potential damage to its traditional relationships within Scandinavia, while the recently finalized Common Fisheries Policy demanded too many concessions on issues of national interests.

Quite apart from the demise of de Gaulle's influence, there were a number of reasons why the EC eventually welcomed the first enlargement. It was, above all, a logical extension of the economic principles which had led to the establishment of the customs union and the common external tariff. A Community of Nine would be more powerful and more competitive than a Community of Six. In addition, while Ireland was relatively underdeveloped, the economic structures of Denmark and the UK were broadly in harmony with those of the existing members, so that integration should have been relatively painless for the Community. The enlargement was also politically important and was seen as a necessary condition if the EC was to become a global superpower.

The UK, and its Commonwealth food suppliers, had to pay a price for its EC membership: acceptance of the CAP, a large budgetary contribution, and a worsening of the balance of payments, at least in the short term (Haack 1973). The static effects of membership were anticipated to be negative, amounting to a reduction of 0.75 per cent to 3 per cent in GDP. In compensation, there were expectations of positive dynamic effects, but these would not follow automatically. The new members did not have to accept the *acquis communautaire* immediately but were given transition periods. The UK had a five-year transition for the CAP so as to offer temporary protection to its traditional Commonwealth suppliers.

In practice, the first enlargement was to prove troublesome, but in January 1973 it appeared to be the successful culmination of the first stage in the construction and evolution of the EC. This was enhanced by the formalization and strengthening of ties with the rump of EFTA. As of 1974, a free trade agreement came into force between the EC and EFTA, thereby ending the arbitrary division of Western Europe into two different trade regions, leastways for industrial goods. This confirmed institutionally that the area was 'a single economic space' (Wijkman 1989, 12). Western Europe had recovered from the effects of the Second World War and from the disruption of ties with Eastern Europe. The EC, as the core of Western Europe, seemed poised in the early 1970s to develop into a major economic and political force in its own right.

FOUR

Global Crisis, Political Paralysis and New Challenges: 1973–85

Bright Openings and Bitter Endings

The early 1970s were years of considerable optimism in the EC. The Community had prospered economically, it had become increasingly integrated in terms of trade and investment, and a number of common policies were in operation. The first enlargement had been agreed and the inclusion of the UK would make the EC the undisputedly dominant economic group in Western Europe. Furthermore, there was a rising tide of opinion in favour of advancing economic and political integration. The 1969 Hague Summit had expressed a general commitment to this goal, while the influential 1970 Werner Report had advocated a three-stage transition to political and economic union. The culmination of this wave of optimism was probably in 1972 at the first summit of the heads of governments of the Community of Six plus the Three-in-waiting. The communiqué of this meeting committed the future EC9 to a much fuller degree of economic integration by the end of the decade: the completion of the internal market, common tariffs, a common currency and a central bank were on the agenda.

However, the optimism of this summit was soon dispelled as the global economy slid into crisis and recession, and political harmony was replaced by political discord. One of the critical factors was the first oil crisis in 1973–4. Edward Heath (1988, 200), the British Prime Minister who had taken the UK into the EC, was later to comment that 'After the oil crisis of 1973–4 the Community lost its momentum and, what was worse, lost the philosophy of Jean

Monnet: that the Community exists to find common solutions to common problems.'

As the following discussion demonstrates, there was no shortage of common problems in this period. However, there was a marked failure to devise and agree common solutions. It became clear that the significant advances of the first period of the Community's operations had been greatly facilitated by the exceptional global economic conditions of those years. As Hodges (1981, 6–7) stated: 'the Community had been "condemned to succeed" in its phase of negative integration because there was nothing to stop it.' More positive policies were required in the 1970s and the 1980s, but by then global economic conditions were making governments more protective, and the road to integration had become far more difficult.

The political context for integration also changed during the 1970s, especially after the resignation of President de Gaulle in 1969. France and Germany, which had constantly been at loggerheads during the 1960s, especially over the CAP, grew closer together in the 1970s. President Giscard d'Estaing claimed, in 1977, that this new entente constituted 'the cornerstone of all progress in the construction of Europe' (Story 1981, 188). If there was an unofficial alliance between the two main powers in the EC, it was only as strong as their mutual self-interest. Yet this alliance was probably critical to what little progress there was in integration during these years.

There was not always a communality of interest even between these two powers and, when that nexus failed, discord was the likely outcome. A prime example of this was the CAP where the system of unified prices broke down and was replaced by nationally divergent 'green prices'. It would be wrong, however, to reduce the politics of intergovernmentalism to this single, if central relationship. In reality, there was a complex web of alliances, and these shifted from meeting to meeting of the Council of Ministers, according to the issues and the national interests at stake. In addition, the UK's membership of the Community was to prove highly problematic. Because the UK had a relatively small farming sector, and because it also purchased a large proportion of its food from outside the EC, it had a structural budgetary deficit. It paid substantially more to the Community than it received in terms of Community expenditure. This was to be a constant political sore in the UK's relationships with other member states, and it was to handicap further the endless compromises and minimalist solutions which had become so characteristic of the EC's intergovernmentalism.

This political paralysis coincided with an increasing globalization of both economic and political issues. Oil crises, superpower politics, and trade negotiations all demanded a coherent and co-ordinated response from the EC member states. But the sheer magnitude of events, as well as their internal political needs, severely limited the possibilities of a collective response. The response of the EC was also constrained by its essentially arbitrary construction as an economic and political region within Western Europe, even though this was lessened to some extent by the enlargements of the 1970s and 1980s. As Hoffman states: 'much harm has been done to the Community by the "globalization" of the issues within its jurisdiction – the fact that the half-continent is an inappropriate framework for dealing with energy, with North–South issues or with the Arab–Israeli conflict' (1983, 31).

By the 1980s the EC had been exposed as structurally limited and incapable of providing a coherent Western European response to global events. The optimism of 1957, and even of the 1972 Summit, was giving way to deep pessimism. Dynamism had been replaced by passivity and 'the infant which held so much promise twenty five years ago has changed into a feeble cardiac patient' (Dankert 1982, 3).

Hard Times: Global Economic Crisis

The long post-war boom already seemed to be drawing to a close in the 1960s and both the UK and France recorded the early warning signs of a possible recession. The precise causes of the crisis were complex but involved the break-up of the international monetary system, labour market rigidities, intensified global competition and two major oil price rises. While these impacted differently on the individual states, the EC as a whole was severely affected by what was to become a world crisis. Integration could offer no protection against such a global recession. On the contrary, the dense network of trade and investment flows which had flourished within the EC meant that the effects of the economic downturn would be transmitted even more quickly amongst the member states.

The main reason for the failure of the international monetary system was that the American economy was no longer strong enough to underpin it. During the 1960s there had been a series of large and cumulative American trade deficits, contributed to in part by the improving trade position of the EC. The resulting dollar surpluses had helped to finance the expansion of world trade, but they had also meant a serious erosion of the USA's gold reserves from $18 billion to $11 billion during the decade. Devaluation would have been the classic response to such recurrent trade deficits but this option was not open to the USA while there was a fixed dollar/gold conversion rate. Inevitably, the fixed parity value of gold had to be abandoned (in 1973) and the resulting currency speculation contributed to destabilization of the world economy.

Production conditions were also changing because of the greater power of labour in the market place. Increasing labour militancy translated into higher real wages and from the late 1960s these outstripped productivity gains, thereby contributing to a fall in profits and investment. Some countries, such as FR Germany, were more successful than others, such as Belgium and the UK, in containing labour costs. However, there was a general rise in real wages throughout Western Europe, fuelled by a tightening of labour markets in the late 1960s. Furthermore, when the economic downturn came in the 1970s, labour market rigidity meant that real wage cuts were resisted, and the economic strain was taken by higher prices, lower profits and a slow-down in output growth.

Another important change in the 1970s had been an intensification of global competition, at first from Japan which had a $10 billion trade surplus with the EC as early as 1980. Later there was intense competition from the Newly Industrializing Countries

(NICs): in 1980 the EC had a 2.9 billion ECU deficit in its trade with the Asian NICs, but by 1987 this had deteriorated to 6.7 billion ECU (Commission of the European Communities 1989). There was also the emergence of competition from Eastern Europe, although the Iron Curtain offered some protection to the EC. The competition sometimes took the form of 'dumping' by countries which were desperate for hard-currency earnings, but EC and other Western transnational companies also bought-back from joint ventures on the other side of the Iron Curtain. However, the most significant competitive threat came from Japan, which was emerging from a period of highly protected growth and restructuring to become a major trading force. By the 1980s it was to overshadow both American and European production in many important medium- and high-technology sectors.

Rapid industrial growth was already evident in the NICs in the 1960s, but it was in the 1970s that they really began to out-perform the developed countries, including the EC. Whereas the annual growth of manufacturing in the advanced economies fell from 6.2 per cent to 3.0 per cent between the 1960s and the 1970s, that of the NICs fell only from 7.3 per cent to 5.8 per cent (Ballance and Sinclair 1983). The expansion of the Pacific Ring and of Latin America was especially impressive; between 1973 and 1979 the exports of manufactured goods from South Korea, Taiwan, Hong Kong, Singapore and Brazil alone increased in value from $18 billion to $53 billion (Jenkins 1984). This has led to import penetration in Western Europe and to the loss of markets for EC transnationals in the Third World. As a result, EC companies have found it difficult to compete in a range of low- and medium-technology industries, especially where labour costs are critical, as in clothing manufacturing.

A crisis of overproduction was already looming in the global economy in the early 1970s, but it was to become critical as a result of two oil price crises. Following the 1973 Israel–Arab war, there was an Arab oil embargo which led to almost a fourfold increase in prices. Price inflation and falls in output followed throughout the developed world. The EC was more affected than Japan in this crisis, partly because real wages fell in the latter while profits and investment were sustained. Arguably, this is when Japan began to secure a decisive competitive advantage over the EC. The second oil crisis, in 1979–80, followed the oubreak of the Iran–Iraq war; within five months of December 1979, oil prices had risen from $12 to $31 a barrel. This time the effects on the developed economies were even more severe and a 3.3 per cent growth rate in 1979 became a −0.3 per cent decline in 1982. The EC was particularly adversely affected because FR Germany – the central motor of growth – had been severely depressed by this second oil crisis. The EC, consequently, recovered more slowly from the crisis than either of its main rivals, the USA and Japan.

The relative weakening of Europe's position is clearly evident in unemployment rates (figure 4.1a). Until 1980 the EC's unemployment rate had been substantially below that of the USA. After 1979 unemployment rates rose sharply in both the USA and the EC and were three times higher than in Japan. However, while American unemployment rates were to fall markedly after 1983, those in the EC remained stubbornly high throughout the mid-1980s. By 1993 there was a 3 percentage point gap between the two countries.

The EC's aggregate growth rate also lagged behind those of its major rivals. For example, in 1979–82 the annual average GDP growth rate in the EC12 was only 0.6 per cent, which was substantially less than the 3.7 per cent in Japan even if it was greater than the −0.3 per cent recorded in the USA (table 4.1). However, the position of the EC worsened significantly in the period 1982–7: its rate of growth was only 2.2 per cent compared to 3.6 per cent and 3.7 per cent, respectively, in the USA and Japan. The EC also lagged behind the non-OECD countries, especially the Asian NICs. This is reflected in employment growth, with the EC lagging behind Japan, the USA and the Pacific Rim NICs (Oceanic) (figure 4.1b).

There were also worrying signs in the global trade statistics. Europe's recovery of its position as the dominant world trading bloc had been one of the most notable features of the 1950s and 1960s.

a) Unemployment rates (%)

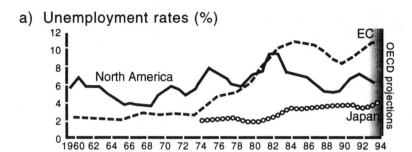

b) Employment growth

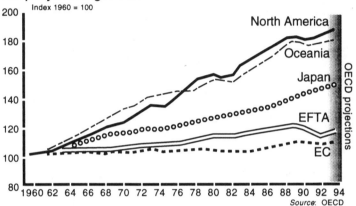

Source: OECD

c) Unemployment (%) by individual countries, April 1992

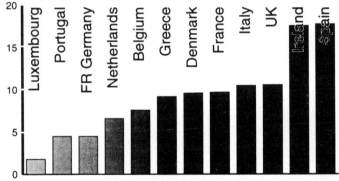

Figure 4.1   Unemployment and employment growth in the EC and major world regions, 1960–94

Sources: Commission of the European Communities (1989); OECD, unpublished data

Table 4.1   World GDP growth rates, 1979–82 and 1982–7

	(Average annual % change)	
	1979–82	1982–87
Germany	0.2	2.2
Spain	0.8	2.7
France	1.7	1.6
Italy	1.2	2.5
United Kingdom	− 0.8	3.1
EC(12)	0.6	2.2
USA	− 0.3	3.6
Japan	3.7	3.7
Total industrialized countries	0.7	2.9
Africa	2.1	0.8
Middle East	− 1.4	− 0.2
America (excl USA and Canada)	1.7	2.4
Asia	5.3	6.9
Eastern Europe	2.7	3.1
Total non-OECD countries	2.6	3.5
Total	1.4	3.1

Source: Commission of the European Communities (1989)

But after 1970 the EC's share of world trade declined from 22.4 per cent to 20 per cent in 1987, while the shares of both Japan and the USA increased (table 4.2). Furthermore, there was growing import penetration in several major industrial sectors. For example, between 1973 and 1982 there was a 19.1 per cent increase in import penetration of the EC10 market for electrical and electronic equipment, which was substantially greater than import penetration in the USA and Japan (Commission of the European Communities

Table 4.2   World trade, 1970–91

% trade accounted for by:	1970	1980	1987	1991
EC12	22.4	21.2	20.0	21.8
USA	16.1	14.1	16.8	16.5
Japan	7.6	8.3	9.6	9.8
Others	53.9	56.4	53.6	51.9

Source: EUROSTAT Key Figures, Supplement, Target 92, no. 2 (1989);
Eurostat Rapid Reports on Foreign Trade

1988c). Some of the most dramatic changes occurred in the car industry; EC car exports fell 23 per cent between 1970 and 1980, at a time when world car exports increased 426 per cent (Ziebura 1982, 134). In the face of such challenges the member states retreated into protectionism. Consequently, the EC which, with the USA, had been a major driving-force for industrial trade liberalization in earlier decades, began to implement protectionist legislation in a number of industries, including textiles, cars and steel (Ziebura 1982).

Above all, the EC seemed to lack flexibility in its capacity to respond to the increasingly rapid pace of change in global economic conditions. In part, this was because the economic crisis exposed some of the limitations of intergovernmentalism. There were increasing difficulties in agreeing common policies and common positions which met the interests of an increasingly diverse set of European states. The national economies of the EC occupied very different positions in the international division of labour, particularly after the 1986 enlargement. At one extreme was FR Germany, which was the only truly global economic superpower in the EC, while on the other hand economies such as Greece and Portugal were not very different, structurally, from some of the NICs. As a result, 'the Community finds itself almost totally powerless, as each member state sees itself confronted with specific problems arising from its economy's own particular position in the international division of labour' (Ziebura 1982, 129).

The New Economic Geography of the European Community

The formation of the EC undoubtedly contributed to the strengthening of trade between the member states. Between 1958 and 1980, intra-EC trade increased twenty-three times while EC trade with the rest of the world expanded only elevenfold (Ziebura 1982, 132). This progressive integration of world trade is shown in figure 4.2; between 1970 and 1990 the proportion of intra- as opposed to extra-EC trade increased from 51.8 per cent to 60.0 per cent. While this can be seen as the positive outcome of trade integration within the common external tariff, it can also be viewed as a failure to compete effectively outside this protected zone. However, even if viewed positively, these figures do conceal a weakening of integration in some sectors. There was in fact a decline in the proportion of intra-EC trade in manufacturing during these years (Dunning and Robson 1987). This reflected the growing import penetration achieved by Japan and the NICs in the markets for products such as cars and televisions.

The levels of trade integration have not always risen constantly within the Community. Indeed, the first enlargement led to a reduction in intra-trade within the EC9 during the mid-1970s, compared to levels in the EC6 (Reichenbach 1980). However, in the longer term, the size and dynamism of the enlarged Community did lead to a major reorganization of trade within Western Europe between the two main trading blocs. By 1986 all the EC9 member states had stronger trade links with the EC than with EFTA (see table 4.3). This applied even to Denmark which had strong traditional trading links with the other Scandinavian countries. At the heart of the EC was a group of major industrial economies – FR Germany, France, Italy, Belgium and the Netherlands – which had almost one-half, or more, of their trade locked into the EC. Comparisons with the position in 1958 (table 3.2) confirm the magnitude of trade integration within this group.

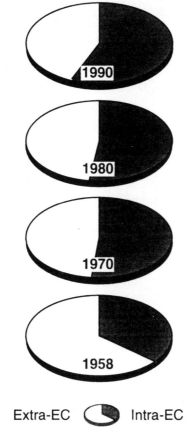

Extra-EC   Intra-EC

Figure 4.2   Intra-EC and extra-EC trade, 1958–90
Source: Eurostat

The first enlargement – and especially the departure of the UK and Denmark – also underlined the diminishing viability of EFTA. By the mid-1980s all the non-EC9 economies, with the exception of Sweden, had stronger links with the EC than with EFTA. Indeed, more than half the exports of Switzerland, Austria and Portugal were directed at the EC. Therefore, there was still a mismatch between the boundaries of the EC and of Western Europe as a

Table 4.3   Trade matrix for Western Europe, 1986

	% manufacturing exports of individual countries to the EC9 and EFTA5	
From	To	To
EC group	EC9	EFTA5
Denmark	37.5	31.0
UK	38.8	8.7
FR Germany	47.5	17.5
Italy	51.4	9.9
France	52.4	8.6
Netherlands	69.1	8.4
Belgium	71.9	6.5
Non-EC group		
Sweden	16.8	21.1
Finland	33.5	22.7
Norway	45.3	18.6
Switzerland	53.8	7.5
Austria	59.4	12.8
Portugal	66.4	14.5

Source: after Wijkman (1989)

functional economic zone. It also helps explain the pressures for further enlargements of the EC in the 1980s and 1990s.

While the EC was beset by economic and political difficulties during these years, it still continued to expand and to prosper, at least in aggregate terms. However, not all the member states benefited equally from the creation of a customs union and the common market. There were considerable variations in growth rates between 1973 and 1983 (table 4.4). Only France, FR Germany, Italy and Ireland experienced growth rates in excess of 2 per cent, while the UK trailed behind with only 1.1 per cent. Interestingly, these data also show that the range of growth rates was as great amongst non-EC countries as amongst member states. EC membership was clearly neither a panacea for national economic ills nor a formula for

Table 4.4   GNP per capita growth rates, 1973–83
(average annual % growth rates)

	Percentage change	Non-EC	Percentage change
EC9			
Belgium	1.8	Austria	2.8
Denmark	1.8	Finland	2.7
France	2.5	Norway	3.7
FR Germany	2.1	Sweden	1.3
Ireland	3.2	Switzerland	0.7
Italy	2.2		
Luxembourg	NA		
Netherlands	1.5		
UK	1.1		
EC12 additional members			
Greece	3.0		
Portugal	NA		
Spain	1.8		

Source: World Development Report 1985, New York: World Bank

guaranteed economic expansion. National economic performances were, undoubtedly, conditioned by EC membership but they were also dependent on domestic circumstances and on global economic movements.

FR Germany had the strongest economic performance, experiencing a long and sustained economic boom from the late 1940s which was broken only by the impact of the second oil crisis. This was facilitated by a national economic strategy of relentlessly pursuing low inflation and a positive balance of payments, combined with a 'social-market' framework for capital–labour–state relations. It became the driving force of EC development, not least because of its very high propensity to import and export (Emminger 1981), and because of its sheer size. In 1980 it accounted for 25.7 per cent and 26.6 per cent of intra-EC imports and exports, respectively. Its nearest rival, France, accounted for only 18.0 per cent and 16.3 per

Number of unemployed as a % of the civilian labour force

Figure 4.3   Unemployment in the EC, 1985
Source: Commission of the European Communities (1989)

cent, respectively, of the totals (Ziebura 1982). The economic strength of FR Germany was also revealed in its low unemployment rate of 8.4 per cent in 1985, compared to an EC average of 11.6 per cent (figure 4.3). The contrast with individual member states was even greater, for Ireland, Spain, Belgium and the Netherlands had rates in excess of 13 per cent. These data also confirm that by the 1980s unemployment had become a structural feature of almost all EC economies, in marked contrast with the USA and, especially, Japan.

The existence of the EC also contributed to the continuing internationalization of capital during these years (Dunning and Robson 1987, 111). While markets for capital and financial services were still mostly nationally organized, businesses could not ignore the importance of the EC. This contributed to both the expansion of international trade and inter-firm trade within the EC. European, American and Japanese companies were all involved in different ways in this internationalization of production and distribution within the Community. For example, in 1983–4 Western Europe's 1000 largest companies were involved in 29 intra-EC and 25 extra-EC acquisitions and mergers (*Financial Times* 5 October 1987), in addition to their investments in greenfield sites and the expansion of existing plant. However, it is important not to exaggerate the extent of internationalization. The activities of most large European companies were still concentrated in only one or two countries. Thus the 1000 companies referred to above had made 101 national acquisitions and mergers in 1983–4, which was twice the number of international deals.

The USA and Japan were the two most significant sources of inward investment to the EC in this period, although Swiss and Swedish companies were also important. As in the previous period, American companies continued to diversify away from the UK and they sought to refine their European strategies. By 1985 they had cumulative investments of $106.8 billion in Western Europe. The UK was still the main location with $33.96 billion but it was followed by FR Germany with $16.75 billion, and then France and the Netherlands with over $7 billion each. It is likely that accession to the EC was critical in allowing the UK to remain the main destination of inward investment. Japanese investment was comparatively new to Western Europe and only amounted to a cumulative total of $11 billion in 1985. The UK held pole position with $3.14 billion. However, it was closely followed by the Netherlands, FR Germany and Luxembourg, each of which had more than $1 billion. The very existence of the EC meant that Japanese companies took a more trans-European view than had earlier American investors. However, the UK was still the most attractive location because of the advantages of the City of London, the use of English as the language of international business, and a national open-door policy for inward investment.

The EC was more than just a passive recipient of inward investment, for it was also home to many of the world's largest transnational companies. Between 1962 and 1982 Western Europe's

Table 4.5   Global distribution of transnational companies,
1962 and 1982

	Percentage distribution of sales	
	1962	1982
USA	67.3	47.8
Western Europe	26.9	31.9
Japan	3.6	12.8
Rest of World	2.1	7.5

Source: Dunning and Pearce (1985)

share of the transnationals increased from 26.9 per cent to 31.9 per cent at a time when the American share fell in the face of Japanese expansion (table 4.5). FR Germany was again dominant and in 1983 it was the country of origin of seven transnationals with sales in excess of $10 billion. This was closely followed by France with four such transnationals and the UK with three, plus two jointly hosted by the Netherlands (Dunning and Pearce 1985). As in the 1950s and the 1960s, they were involved in building up trans-European networks, partly in response to the creation of the EC.

However, many EC-based transnationals have disinvested, in relative terms, from Western Europe in response to rising production costs, increased NIC competition, and uncertain demand following the oil crises. Direct production in, or sub-contracting to firms in, the less-developed countries offered significant cost reductions. EC transnationals have also been attracted to invest in the USA which is the world's largest single market. Therefore, there have been two parallel processes in operation: greater integration of capital within the EC and between it and the rest of the world. In other words, new divisions of labour have been created both within the EC and at the global scale. The economic crisis of the 1970s served to intensify this reorganization of capital.

The main effect of the economic crisis on the common labour market was to reduce the numbers of international migrant workers in the EC9 from 6.6 million in 1973 to 4.3 million in 1985 (Molle and

Van Mourik 1988, 326). At this latter date there were 1.2 million migrants from the EC9, 0.8 million from Greece, Spain and Portugal, and 2.3 million from the remainder of the world. As in the 1960s, Mediterranean countries, such as Turkey, and ex-colonies continued to be the main sources of such migrants.

On balance, it seems that the creation of the common market for labour had had little discernible effect on the ratio of EC to non-EC international migrant workers in the Community. In fact, the percentage decline in the numbers of EC international migrant workers in the Community was twice as great as the rate of decline of non-EC migrants during the most severe crisis period, 1980–5 (Molle and Van Mourik 1988). This is surprising in that the differential legal access of EC migrants should have been more significant at a time of economic recession.

Member states could seek to limit the numbers of non-EC immigrants but could not interfere with the freedom of movement guaranteed to EC citizens by the Treaty of Rome. Measures were taken to restrict the entry of non-EC nationals. For example, there were simple bans on immigration, such as France, Belgium and FR Germany introduced in 1973–4. There were also incentive schemes to encourage existing migrants to return home; France paid 10,000 francs per capita to such migrants under a scheme introduced in 1977, and FR Germany paid subsidies to assist returnees to establish small businesses or to obtain essential training. In practice, however, the main effects of the bans were to encourage existing migrants to stay longer, even though job markets were tightening significantly; few would risk returning home early, given that re-entry could be difficult. Nor did the legislation end the inflows from non-EC countries; it only reduced them. This was because the process of family reunification was allowed to continue unrestricted in most countries except the UK (Castles et al. 1984). International migration, ultimately, was controlled more by economic and social conditions than by legal access – at least in the short and medium terms. The common labour market was operating as an integral part of, rather than in isolation from, global labour markets.

The Second Enlargement: the Mediterranean and the European Community

Although the EC was beset with many problems, its size, relative dynamism and increasing dominance of European trade meant that there were pressures from potential applicants for a second enlargement. The most likely candidates were three Southern European states – Greece, Spain and Portugal. Greece had held associate membership since 1962, while Portugal and Spain had signed special trade agreements in 1972 and 1970 respectively. These involved limited trade liberalization and the transfer of some development funds to Europe's relatively poor southern periphery. However, all three were barred from full membership by virtue of their non-democratic governments. This constraint was to disappear in dramatic fashion in the mid-1970s. In 1974 and 1975 dictatorial governments were overthrown in all three countries and, by 1977, had been replaced by elected parliaments and democratic constitutions (see Williams 1984). Enlargement of what can be called the European 'democratic space' was a precondition for the enlargement of the economic and political space occupied by the European Community.

When the Southern enlargement first appeared on the agenda in the mid-1970s, this did not seem to pose any great difficulties for the Community (Seers 1982). The EC was recovering from the immediate impact of the 1973–4 oil crisis, and the creation of a larger market was seen as a logical economic goal. Greece applied for membership in 1975, almost immediately after the overthrow of the colonels' regime. The negotiations were relatively painless. Greece has a small economy and, while it would increase competition for existing Mediterranean farm producers (wine, olive oil, citrus fruits and so on) within the EC, it also offered important new markets for industrial goods. There were also political reasons for the EC's rapid acceptance of Greek membership. Osborn (1988, 14) writes that 'There is a widely held belief in Brussels that Greece was technically and administratively unprepared for EEC membership in 1981, when the date of accession was set, in order to secure entry before the then anti-EEC socialists could come to power.' It is true that, in many ways, Greece was unprepared for EC accession. Even a six-year transitional period for industrial goods did not offer sufficient protection to Greek manufacturers. Their lack of competitiveness was exposed and Greek industry was given a two-year extension to the transitional arrangements, so as to soften the impact of accession.

Greek membership of the Community has been a difficult experience for both the Community and for Greece itself. In part, this is because 'in the process of integration, political parties are key actors in so far as they transmit opinions, shape policy choices and participate in decision-making' (Featherstone 1989, 248). The key element, in this context, was the 1981 electoral victory of PASOK (the Greek socialist party) with its populist and strongly anti-capitalist and anti-USA programme (Lyrintzis 1989). The EC was seen as representing an extension of American hegemony into the European space. In addition, PASOK was strongly nationalistic, and one of its electoral slogans was 'Greece for the Greeks' (Featherstone 1989); consequently, it opposed the loss of sover-eignty implied by Community membership. In government, PASOK modified its stance considerably and its rhetoric did not often match its policies. It secured a number of concessions from Brussels in 1983, which led Papandreou to declare:

> From the moment it became evident that Greece's insistence on the necessity for solidarity towards the countries of the European South had been accepted by the wealthier northern partners, we judged the country's interests to be better served inside rather than outside the European Community. (Quoted in Osborn 1988, 14–15)

Solidarity in this instance meant increased assistance to Greece from the Community's structural funds (£7.5 billion in total by 1991) and a 50 per cent allocation from the specially created Integrated Mediterranean Programme (equivalent to £1.4 billion by 1991). Yet, in reality, EC membership has been a very mixed economic experience for Greece. Its trade balance has deteriorated, inflation has risen particularly as a consequence of the introduction of CAP prices, and manufacturing output fell 6 per cent during the first three years. The CAP support system has also been less favourable to Greece than to other member states: only 75 per cent of Greek farm production compared to 95 per cent of average EC farm production is actually covered by the CAP. However, Greece has been a net recipient of EC budgetary transfers; even in 1984 this amounted to 995 million ECU, equivalent to 2.5 per cent of GDP (Georgakopoulos 1986).

The political value of membership has also been questionable. While membership of the EC has given Greece part of a larger European political voice, it has often been at odds with this. For example, Greece did not support the dispatch of European forces to Sinai in October 1981, and did not condone the imposition of commercial sanctions against Poland following the introduction of martial law. Yet both of these issues had been discussed in the forum of European Political Co-operation, the EC's mechanism for common foreign policy (Ioakimidis 1984). Greece has also frequently been isolated at meetings of the Council of Ministers. However, membership has offered political advantages. Much of the increased structural financial assistance negotiated by PASOK was channelled to Greek rural areas. Not surprisingly, PASOK continued to draw strong electoral support from these areas in the 1989 and 1990 Greek elections, at a time when national opinion had turned against the party. It would be no exaggeration to state that PASOK was saved from humiliating electoral defeat by the way it had been able to influence policy-making in Brussels. Greek membership of the Community also had longer-term political implications; it meant that Greece was in a position to influence any future Southern enlargements of the Community. This would be critical in negotiations with Turkey in the late 1980s but, more immediately, it would influence the Iberian enlargement.

Portugal and Spain applied for membership of the EC in 1977, almost immediately after the completion of the formal transition to parliamentary democracy. By the time that negotiations began in earnest, in the late 1970s and early 1980s, conditions within and outside the EC had become more difficult. The 1979–80 oil crisis had severely depressed the European economy and Greek membership was proving difficult to digest. It was also becoming apparent that there were major differences in the economic and social structures of the EC9 and of the three Southern applicants. Not least, they were likely to make considerable demands on the Social and Regional Funds which, realistically, could not be met within existing budgetary constraints. Spain was also a major economy and its absorption was likely to have considerable economic implications for both industrial and agricultural producers within the Community. Faced with these difficulties, and increasingly preoccupied with the need for fundamental reform of its own institutions and policies, the Community had great difficulties with the Iberian applications.

There were two compelling reasons for the Iberian countries to apply for EC membership. Firstly, it was presented in domestic politics as a way of stabilizing the new democratic institutions. Once they were members of the Community, their economic structures would be reshaped by relationships with the other member states. As non-democratic governments could not remain members of the EC, there would be intense economic pressures to prevent any reversion to dictatorial governments. Secondly, it was argued that Spain and Portugal were already effectively members of the EC by virtue of their strong trade, investment and labour migration ties with the Community. This was felt most acutely in Portugal as the UK, traditionally, had been its principal trading partner and source of inward investment. Membership would allow them to participate in EC decision-making which impacted on their economies.

There were strong trade links between Spain and Portugal and the EC. These had been recognized in the special trade agreements signed in the early 1970s. However, contrary to expectations, the trade agreements had not led to a significant increase in trade integration between the EC9 and the two applicants (Guerrero et al. 1989). There had been an increase in the shares of the exports of Spain and Portugal which were destined for the Community (table 4.6). By 1983 these were 57.7 per cent and 47.5 per cent, hence

Table 4.6   Trade between the EC9 and Spain and Portugal, 1970 and 1983

| | | Share of EC9 in trade of Iberian countries | |
		Exports	Imports
Portugal	1970	41.9	48.3
	1983	57.7	39.7
Spain	1970	46.3	40.3
	1983	47.5	32.3
		Share of Iberian countries in trade of EC9	
Portugal	1970	0.9	1.4
	1983	0.8	1.4
Spain	1970	2.1	4.1
	1983	3.5	5.0

Source: Donges and Schatz (1989, 279)

underlining the advantages of membership as a way of guaranteeing access to such important export markets. However, the proportion of Spain's and Portugal's imports originating from the EC actually fell during the 1970s and was less than 40 per cent in both cases by 1983. This was partly due to growing food imports from the Americas and to manufacturing imports from Japan and the NICs. In this case enlargement had a compelling logic for the Community as a way of bringing Spain and Portugal within the protectionism of the common external tariff.

The negotiations were long and difficult, and were complicated both by domestic politics as well as by the Community's need to implement internal institutional reforms. Greece used the negotiations as an opportunity to gain further structural assistance for its own economy, especially from the Integrated Mediterranean Programme. France and Italy were concerned to protect their farmers from the strong competition offered by Spain's agricultural exports. It took eight years of painful negotiations before a treaty of accession was signed in 1985. Spain and Portugal were incorporated into the EC space in January 1986. The main elements of the treaty were as follows:

Agriculture.   Transition periods of up to ten years to protect EC farmers from Iberian (mainly Spanish) competition, especially in the markets for olive oil, vegetables, wine and citrus fruits. At the same time the transition periods gave protection to Iberian farmers from full EC (especially French) competition in the markets for cereals, meat and dairy products.

Fishing.   Spanish and Portuguese membership would increase the size of the EC fishing fleet by 80 per cent. They were, therefore, granted only limited access to other member states' fishing grounds during a ten-year transitional period.

Manufacturing.   There was a seven-year transitional period for the progressive adoption of the common external tariff and the customs union.

Budget.   Transitional arrangements were offered until 1992 but, thereafter, the new members would have to meet their budgetary contributions in full.

In practice, the economic implications of accession for Spain and Portugal largely mirrored the Greek experience. For example, in Portugal between 1985 and 1988 there was a negative trade effect: imports increased 120 per cent while exports increased only 88 per cent. However, this was partly counterbalanced by net EC budgetary transfers to Portugal, equivalent to 1.6 per cent of national GDP. Yet the Iberian enlargement was to prove easier than the Greek for the Community. In part this was because global economic conditions improved in the late 1980s, hence cushioning the impact of the transition. Domestic politics were also more helpful as there had been broad inter-party consensus in Spain and Portugal about the value of Community membership. Additionally, the enlargement coincided with renewed optimism about the future of the Community as the 1992 programme reinvigorated a static decision-making process.

In retrospect, it seems clear that the major difficulties of the second enlargement stemmed more from the internal problems of the EC, and from unfavourable global economic conditions, than from any inherent characteristics of the new members. The very fact that the enlargement was accomplished during this difficult period in the Community's history bears witness to the enduring strength and appeal of the logic of economic and political integration which had inspired the signing of the Treaty of Rome in 1957: trade links, economies of scale, efficiency, and the creation of a strong and independent, collective Western European voice in world affairs. This was recognized by the Commission of the European Community at the time:

> The pattern of expansion of the European Community over the years has been from the centre outwards, first towards the North, and now to the South. The absence of Spain and Portugal, both major trading partners with the EC, for so many years, was an obvious gap in the trading bloc of Western Europe. The addition of two countries with such rich trading links with Africa, the Middle East and Latin America is a major advantage for the existing Community. Finally, there is a political interest in supporting and encouraging the common values of democracy and freedom in Spain and Portugal. For those two countries, the economic and political necessities of today's world made an overwhelming case for Community membership. (1986a, 1)

While the question of the Iberian accession was eventually resolved, this created new difficulties for the EC. By the 1970s the Community had signed trade agreements with all the Mediterranean countries except Albania and Libya; these offered limited agricultural concessions, tariff-free access for most industrial goods and assistance from the European Investment Bank. The accession of Spain and Portugal provided real threats to those Mediterranean countries which were still outside the EC. Not least, industrial and agricultural exports from Spain to the EC exceeded those of all the other Mediterranean countries taken together – and, as of 1986,

Spain would have privileged 'insider' access to this important market. To alleviate these concerns, new trading agreements were signed with countries such as Algeria and Egypt. However, relationships between the EC and the Mediterranean countries remain open to further revisions, including the possibilities of further accessions (see pp. 219–20). Turkey, in particular, has felt disadvantaged by the Mediterranean enlargements of the Community, its sense of exclusion being heightened as it had enjoyed a relatively favoured position (along with Greece) in the 1960s because of its early Association agreement with the Community (Balkir and Williams 1993).

The second enlargement was agreed in 1985 while the Community was still struggling to agree essential internal political and economic reforms. There were also continuing difficulties over the budget and in advancing the corpus of common policies; this is the subject of the final section of this chapter.

Common Policies and Common Failings

The capacity of the Community to function as more than a customs union and a series of common markets depended on the development of common policies which were applied throughout the EC space. In this respect the 1970s and early 1980s were a disappointing period in the Community's history. As Wallace stated:

> the rumbling and linked disputes over the distributional consequences of the Community budget, the reform of the CAP, and the application for membership by Spain, have demonstrated a worrying rigidity in structure; the Community's capacity to adapt its policies and institutions to changing circumstances appears to be low. (1982, 63)

As ever, the most successful but the most problematic area of common policy was the Common Agricultural Policy. The

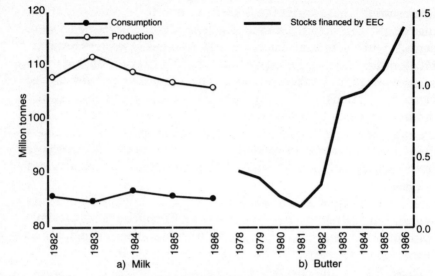

Figure 4.4   The EC dairy surplus, 1978–86
Source: Financial Times, 3 September 1986

fundamental weakness of the CAP was the combination of its particular form of interventionism (based on buying up surplus production for storage) with the high levels of guarantee prices which had been agreed as part of the political compromise necessary to establish the farm policy in the 1960s (see pp. 47–9). EC guarantee prices had been set at levels well above those prevailing on world markets, resulting in the costly purchasing and storage of unsold stocks. In the 1970s, there had been some convergence between EC and world prices due to global supply reductions. However, by the early 1980s the gap had again widened. EC production remained substantially above EC consumption and the levels of food stocks increased sharply (figure 4.4). The costs of the CAP rose inexorably and dominated the budget throughout these years. The maximum proportion of the budget devoted to the CAP was 78.4 per cent in 1978 but, even as late as 1985, it amounted to 69.9 per cent (Harrop 1989, 158–9).

Despite these budgetary problems, it could be claimed that the

CAP was a Community-wide set of common policies. However, even this claim was undermined in two important ways during this period. Firstly, the second enlargement brought three major Mediterranean producers into the Community. As the CAP had been formulated by the original Six, it is not surprising that it favoured temperate northern products while many Mediterranean products were excluded or received limited support. Consequently, the accession of Spain, Portugal and Greece weakened the claim of the CAP to be an all-embracing set of policies for farming throughout the EC space.

The second source of difficulty stemmed from the break-up of the international monetary system in the late 1960s. Under the Bretton Woods system of fixed exchange rates, it had been relatively simple to apply a single system of guarantee prices throughout the Community, once their levels had been agreed by the Agricultural Ministers. With the collapse of the Bretton Woods system, floating exchange rates meant that there would be fluctuations in the prices paid to farmers or by consumers. As early as 1969, there was a decisive break following a 12.5 per cent French franc devaluation and a 9 per cent revaluation of the Deutschmark. If CAP prices were converted into national prices at these new exchange rates, there would have been higher prices for French consumers and lower prices for German farmers. As neither outcome was acceptable in terms of domestic politics, both countries were allowed to create artificial 'green' exchange rates for farm prices. These were, respectively, lower and higher in France and FR Germany than they would have been under the new exchange rates (Strauss 1983).

Having created 'green' exchange rates, it was then necessary to establish Monetary Compensation Amounts (MCAs) to prevent serious trade distortion arising from the existence of differential prices within the EC. Positive MCAs involve levies on imports and subsidies on exports in those member states with green prices above nominal EC rates. Negative MCAs imply levies on exports and subsidies on imports for those member states with green prices below nominal EC levels. MCAs were initially conceived of as temporary measures to overcome short-term difficulties caused by 'price breaks' within what should have been a unified price field. However, the complete collapse of fixed international exchange rates in the early 1970s, combined with the value of green prices to national governments for domestic political purposes, meant that they quickly became an established part of the CAP.

In practice, the MCAs did not eliminate all trade distortion arising from differential prices. Strauss (1983) considers that, in the 1970s, they contributed to an undue increase in FR Germany's share of EC farm trade, even though a high-cost producer. At the same time this limited the expansion of output in lower-cost member states. This was to be a constant source of friction in the otherwise generally harmonious political partnership between France and FR Germany. France opposed the use of MCAs throughout the 1970s and the 1980s, arguing that they offered unfair advantages to German farmers, hence disrupting the mechanisms of free trade.

While agriculture held centre-stage in the political and policy deliberations of the EC, there was some, if difficult, progress in other common policy areas:

- The *Common Fisheries Policy* had been agreed in the early 1970s, in advance of the first enlargement. Its main impact was to reduce the exclusive zones reserved for national fishing fleets. For example, the UK was only allowed to reserve one-third of its twelve-mile-zone for its own fishing fleet. Following the 1975 Law of the Sea Conference, the EC adopted a 200-mile exclusive fishing zone, thereby establishing a 'Community Pond'. However, the allocation of common vs. national fishing rights within this zone was, politically, highly contentious. Agreement was, eventually, reached in 1983. All states had exclusive rights within the inner six miles and these were usually extended over the first twelve miles, except where 'historic rights' required selective access for other countries. Member states were allocated specific national quotas within the 12–200-mile zone, and these, inevitably, have been the source of further political difficulties. This delicate balance was upset by the second enlargement of the Community. As a compromise, Spain and

Portugal were given limited access to the Community Pond over a ten-year transition period, hence requiring a major renegotiation in the mid-1990s.

- The *European Social Fund* was given a higher priority following a series of deepening economic crises in the 1970s and 1980s. Its share of the total budget increased over time and had reached 7 per cent by 1986 (table 4.7). There was consensus that unemployment could no longer be seen as an expression of temporary adjustments in labour markets. Instead, there was growing structural unemployment in the Community. Hence, in 1970 the Social Fund was given the tasks of ameliorating lasting unemployment problems in the economy and of assisting those with special employment difficulties such as disabled persons, older workers, women and young adults. Although conceived of primarily as a redistributive social measure, the Social Fund also had an economic function. It helped to refine labour markets (via mobility allowances and training) to meet the needs of capital.
- Although the general idea of a *European Regional Development Fund* was laid down in the Treaty of Rome, it only came into existence in January 1975. There was growing concern that competition, recession and the increasing geographical mobility of capital and labour could lead to greater regional inequalities within the EC. This could further undermine cohesion and harmony in the Community. However, the first enlargement was the decisive event leading to the establishment of the ERDF, for it was seen as a redistributional mechanism to offset the UK's anticipated budget deficit. Initially, there were plans to make the ERDF the dominant (over national programmes) regional policy in the EC. These had to be abandoned, mainly because of German opposition to the high costs. Instead, a more modest ERDF came into existence as a complement to national regional policies. While the Council of Ministers agreed the quota of funds for each member state, there was no attempt to harmonize regional policies across the Community space. Nevertheless, the persistence of regional inequalities in the Community, linked to the accession of less-developed economies in the 1980s, has led to

Table 4.7   EC budgetary expenditure, 1973 and 1986

	1973	1986
ECSC	1.8	1.2
European Development Fund	4.3	2.2
EAGGF (Agriculture)	76.3	63.4
Social Fund	5.5	7.0
Regional Fund	–	6.5
Industry, Energy, Research	1.4	2.1
Administration	10.8	17.6
Total	100.0	100.0
Absolute totals (million ECU)	4,937	36,413

increased funding of the ERDF; by 1986 it received 6.5 per cent of the total budget (table 4.7).

- *Industrial policy* was probably the largest single failing of the Community in this period. After the major achievements of establishing the customs union and the common external tariff in the first period, there was a failure to make further progress. This was even more significant, given the deteriorating trade position of EC industry and the expansion of medium- and high-technology imports from Japan. Virtually the only EC collective response was a retreat into protectionism, one notable example being the 1978 Multifibre Agreement to reduce EC textile imports from leading NIC producers. Positive interventionism was proposed, and there was a scheme for EC participation in the process of industrial restructuring. This was opposed by FR Germany which, significantly, had a large trade surplus for manufacturing goods, and which was fundamentally opposed to industrial interventionism. Once again, the tension between domestic politics and the needs of the Community as a whole was evident. Only the ECSC was involved in effective restructuring strategies, and then only in response to the declaration of a 'manifest crisis' in the steel industry in 1980. The ECSC strategy was to set mandatory national production quotas and to reduce state subsidies (Williams 1987, 185). There was also a failure to develop a common policy towards transnational companies, even though 'multinational and international capital from Europe, America and Japan has managed to help disintegrate sectors of European industry by integrating its own operations world-wide' (Holland 1980, 27).

Whatever progress was made with common policies, the broader move towards greater European integration had stalled in this period. The 1969 Hague Summit, the 1970 Werner Report and the 1972 Paris Summit had all committed the EC to increased economic and monetary union (see p. 57). However, the only concrete achievement of the early 1970s was the establishment in 1972 of the 'snake' of EC currencies. These were to be linked together so as to fluctuate within narrower margins than existed for the 'tunnel' of major world currencies. Even this proved difficult to sustain in the face of changing global economic conditions. Speculation drove sterling and the Irish pound from the snake almost immediately, followed by the Italian lira in 1973. France also had to withdraw temporarily from the snake on two occasions and, by the late 1970s, this system of currency exchange regulation had broken down.

The *European Monetary System (EMS)*, introduced in 1979, was to have more lasting significance. The aim, once again, was to create a single zone of monetary stability within the EC space. It introduced two new features: the European Currency Unit (ECU) as a financial unit of accounting, and the Exchange Rate Mechanism (ERM). EC currencies could move against each other within bands of -2.25 per cent to $+2.25$ per cent, while floating together against other currencies on world markets. Occasional realignments were possible amongst EC currency exchange rates, but these were to be minimized by the co-ordination of monetary and other policies. The EMS was inspired by the German *Bundesbank* and this was one more indicator of the growing dominance of FR Germany within the EC. The central currency in the ERM was the German Deutschmark which was thereby given a historic opportunity to assume its undoubted position as a major world currency.

In practice, the ERM performed less than perfectly. The UK decided not to join from the outset, while Italy had to be given wider margins of −6 per cent to +6 per cent for the lira. There were a series of realignments between the currencies, with seven being necessary in the first four years alone. Yet, the EMS did represent one of the few important major policy advances in this period. According to McDonald and Zis (1989), it had four major achievements: greater monetary co-ordination, convergence of inflation rates, reduced volatility in exchange rates, and improved stability for ERM currencies compared to other currencies. To some extent these successes were fortuitous, for the introduction of the ERM coincided with a strengthening of the dollar on world markets and reduced pressure on the Deutschmark. There was also a common commitment amongst the member states to make monetary policy and the fight against inflation their primary economic goal in the early 1980s (Cobham 1989). Despite these qualifications, the EMS did mark a major step in economic integration within the EC.

Quite apart from external pressures, there were two major internal constraints on the capacity of the Community to advance integration and to develop common policies. The first of these, the CAP-dominance of the budget, has already been referred to, and is confirmed by table 4.7. In addition, there was constant political tension stemming from the UK's recurrent large net contribution to the EC budget. This was a structural deficit arising from high UK food import levies and the small size of British agriculture. In practice, it meant that a whole series of meetings of the Council of Ministers in the 1980s were dominated by the UK's repeated demands for budget rebates. This contributed to British obstructionism on a number of issues, and generally hindered the ability of the Community to address the major issues of integration.

A temporary UK budget rebate was arranged in 1982, but the issue was only settled two years later. Agreement was almost secured at the Brussels summit in March 1984. There had been broad agreement on some reforms of the CAP (for example, milk quotas), on increasing the ceiling on the EC's own resources from 1 per cent to 1.4 per cent of VAT receipts, and on the principles of the budgetary adjustment. All that remained to be determined was the actual size of the UK rebate in 1984. The UK demanded 1.35 billion ECU and was offered 1.1 billion ECU. Mrs Thatcher would not compromise and, eventually, Chancellor Kohl of Germany stormed out of the Council meeting. Taylor argues that this was the decisive turning-point in the negotiations and, indeed, in European integration in the 1980s:

> When he rose from the table in anger at 6.30pm on 20 March 1984 in Brussels, he revealed a carelessness about UK reactions which carried an unmistakable and ominous warning for the latter. They could, as it were, 'take it or leave it'. Kohl's action was decisive in revealing the flaw in the British diplomatic armament – that they were prepared to make sacrifices in order to avoid being excluded from the inner sanctum. (1989, 7)

The Community had been taken to the brink of disaster and the UK had been firmly told that, if necessary, the EC – or at least its inner core – would move ahead without it. Frantic negotiations followed and at the Fontainebleau Summit, in June 1984, an

Table 4.8   EC budget transfers in 1985

	GDP per head as % of EC(12) average	(million ECU) before British rebate	after British rebate
Denmark	153	+ 400	+ 300
Luxembourg	142	+ 300	+ 300
FR Germany	138	− 3,100	− 3,500
France	126	+500	0
Netherlands	112	+ 500	+ 400
Belgium	108	+ 600	+ 500
UK	91	− 3,000	− 1,000
Italy	86	+ 800	+ 1,200
Ireland	66	+ 1,200	+ 1,100
Greece	38	+ 1,400	+ 1,400

Source: The Economist, 20 June 1987 (in Harrop 1989, 161)

agreement was concluded. The UK was to receive two-thirds of the difference between its VAT contributions and the EC's share of expenditure in the UK. The outcome is summarized in table 4.8. In 1985 the estimated net UK contribution was reduced from 3000 million ECU to 1000 million ECU, with France, FR Germany and Italy accepting the largest net increases in their payments. One important consequence was to make FR Germany the dominant financier of the EC, which would increase its power in future negotiations on the development of the Community. The way had also been opened for the advancement of European integration in the mid and late 1980s.

FIVE

Remaking the European Community: the 1980s

Global Relationships and Eurosclerosis

In the 1980s the European Community was faced with a double crisis: loss of global economic competitiveness and the stagnation of political and economic integration. Declining global competitiveness was starkly underlined by the reduction in the EC's share of world trade in manufactured goods from 45 per cent to 36 per cent between 1973 and 1985 (Price 1988). At the same time, the EC was faced with a number of internal conflicts: how to expand its budget, how to finance the CAP, and how to advance political integration. Furthermore, the EC had reached this crisis point in the mid-1980s beset by institutional weaknesses. It could only take major decisions by paying off all possible interests, that is, it achieved minimal advances in return for maximum concessions to national interests.

There was a crisis of economic self-confidence in the EC and this stemmed from the relatively poor performance of the Community *vis-à-vis* its major competitors, Japan and the USA. In the 1980s both the EC and North America had been thoroughly outperformed by south-eastern and eastern Asia, as can be seen in table 5.1. Of even greater significance was the way that the EC had also fallen behind North America in the 1980s. Indeed, EC growth rates were less than the global average during this period.

GDP per capita growth rates were a sensitive area of EC under-performance. In the late 1970s the rate of growth in the EC had exceeded that in both the USA and Japan. However, the EC recovered more slowly after the 1979 oil crisis. It still had a positive overall trade balance in manufactured goods, but this seemed to be

Table 5.1   GDP growth rates, 1980–9

	(% annual change) 1980–5	1985–9
Developed Market Economies	2.5	3.5
North America	2.8	3.6
EC	1.5	3.3
South-eastern and Eastern Asia	7.2	7.6
World	2.7	4.6

Source: United Nations Statistical Yearbooks

in terminal decline. A positive balance of 65 billion ECU in 1981 had been reduced to 37 billion ECU by 1987 (Commission of the European Communities 1989). Throughout the 1990s, EC growth rates in GDP per capita lagged behind those of its main rivals, Japan and the USA (figure 5.1).

This aggregate performance concealed considerable national variations. The severe structural problems of several EC economies, such as the UK and Belgium, were masked by the strong performance of FR Germany, which had become the world's largest industrial exporter (Williams 1987, chapter one). The 1970s and the 1980s witnessed continued increases in the industrial and trading strength of FR Germany, and this was partly at the root of the economic difficulties experienced by some other EC member states. West Germany had increased its exports to Western Europe without there also being an increase in Europe's share of world trade. In a sense, therefore, the rest of Western Europe paid the economic price for German success (INSEE 1989). France, in particular, suffered in the mid-1980s, not least because it was still heavily dependent on markets such as North Africa and the Middle East which were relatively stagnant after the 1970s.

In part, the relative decline of the EC stemmed from an inability to meet the challenge of the NICs. Between 1980 and 1987 the negative trade balance between the Community and the Asian NICs deteriorated from − 2.9 billion ECU to − 6.7 billion ECU (Com-

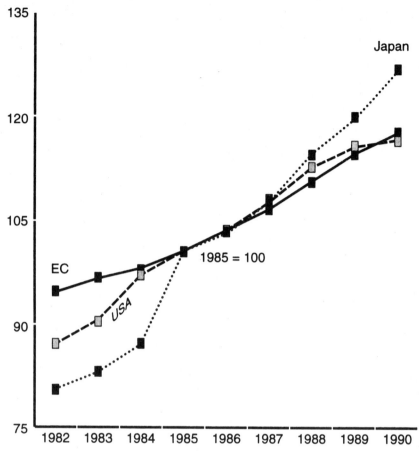

Figure 5.1   GDP per capita volume index, 1982–90, in the EC, Japan
and USA

Source: Eurostat (1993), *Basic Statistics of the Community*

mission of the European Communities 1989). However, as noted
earlier, this represented only a small part of the overall trade deficit
of the EC in manufactured goods. The difficulties of the EC's
position stemmed from sustained import penetration in virtually all
major industrial sectors (figure 5.2). In contrast, import penetration
was considerably less in Japan in all major industrial sectors. It was

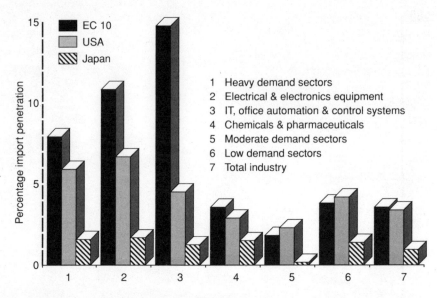

Figure 5.2   Import penetration of home markets for industrial goods in the EC10, USA and Japan in 1982
Source: Commission of the European Communities (1988c)

also less in the USA for all heavy-demand and advanced industrial sectors such as electronics, chemicals and information technology. This gave rise to 'Eurosclerosis', a fear that the EC would be condemned to producing low- and intermediate-technology goods as it was squeezed out of high-technology sectors by Japanese and US competition.

The fears of Eurosclerosis were well-founded. Between 1970 and 1985 the EC had increased its share of world exports of high-technology goods by 1 per cent but it had also increased its share of world imports of such goods by 2 per cent (Geroski and Jacquemin 1985, 176). The most sensitive sector was probably electronics and electrical goods: by 1986 the EC had $83 billion of imports but only $69 billion of exports of these critical high-technology products. The EC was especially weak in the production of computers. EC companies tended to produce for semi-protected national markets and had little penetration of other Community markets, let alone the global marketplace. As a result, the EC computer industry tended to be reliant on the USA for commercial innovation and was slow to react to increasingly rapid changes in the sector. It also meant that EC companies were vulnerable to 'soft' demand in national markets. This was underlined by the problems of Nixdorf, one of Europe's leading computer companies, which experienced severe financial difficulties in 1988–9 following a weakening of demand in Germany. In addition, ICL – the most profitable of mainframe computer manufacturers in the EC – was sold in 1990 to Fujitsu of Japan, as it lacked the R&D capacity to remain competitive in global markets.

The economic weakness of the EC was reflected in the growth of unemployment (see figure 4.1a) which had reached 11 per cent by 1983. Not only was this considerably higher than in Japan and the USA, but there was also a lack of job growth in the EC in the 1980s, compared to its two main global rivals (see figure 4.1b). Unemployment rates in individual countries were even greater and exceeded 20 per cent in both Spain and Ireland. One of the political outcomes of the economic crises was intense political pressure for protectionist measures, both at the national and the EC levels. These took various forms including voluntary export restraints on clothing and textile imports from most developing countries, on steel imports from some NICs, and on consumer electronics from Japan, Korea and Taiwan. The steel industry was especially crisis-prone, and this lead eventually to the 1980 Davignon Plan. This combined production quotas, agreed capacity reductions and protectionism (via voluntary export restraints with the NICs) in an attempt to reorganize the restructuring of a declining EC industry.

Member states also introduced national measures to protect their economic interests. For example, Italy and Spain had virtual embargoes on Japanese car imports, while the UK had Orderly Marketing Arrangements with leading Japanese car producers, limiting their exports to 10 per cent of the UK market. The most infamous instance of protectionism was the 'Poitiers' incident, when France unilaterally announced that all imported video recorders had to be cleared via the relatively remote customs offices at Poitiers; this, inevitably, reduced the volume of imports and added significantly to their costs. Protectionist measures were not simply directed at third-party countries; they were also imposed, with increasing regularity, between member states. This was reflected in the growing number of state-aid and protectionism cases taken to the European Court of Justice: only six cases in 1970 compared to sixty-one in 1981 (Harrop 1989, 94). The result was, once again, to fragment the already fractured economic space of the EC.

Economic relationships also became strained between the EC and the USA. There had been some conflicts of interest between the two trading blocs since the EC first came into existence. However, in the 1950s and the 1960s the USA had been willing to overlook these because of what was perceived to be the political advantages of the formation of the EC. By the 1980s a much weaker USA was no longer prepared to make so many concessions to the EC. Not least, the EC12 was now seen as a major economic competitor; it had a larger market, a larger total GNP and a larger share of world trade than the USA. Behind this lay the deteriorating trade position of the USA: a $25 billion trade surplus with the EC in 1980 had been converted into a trade deficit of $18 billion by 1986. There was acute awareness that FR Germany was considerably outperforming the USA, and could challenge it for global economic leadership (Ginsberg 1989).

Economic friction between the EC and the USA was manifested in relation to a number of issues, but these usually centred on rival claims of unfair protection or subsidies provided for agriculture. For example, the 1985–6 'pasta war' resulted from the US imposition of high import duties on EC pasta in retaliation for the EC's programme of preferential treatment for Mediterranean producers. Another contentious issue was the 1986 Iberian enlargement of the EC. American officials estimated that the terms of accession would have major trade-diversion effects as Iberian food imports became sourced from within the EC rather than from the USA; for corn and sorghum alone the trade losses were estimated as $400 million. Threats of trade sanctions and counter-sanctions flowed back and forth across the Atlantic, before a compromise was reached in 1987 guaranteeing minimum levels of imports from the USA.

Economic divergences also added to a growing political and diplomatic divide between the USA and the EC. This had become obvious during the Yom Kippur War, when the EC had preferred a more even-handed diplomatic approach than had the strongly pro-Israeli Americans. There were several other such instances; for example, the USA considered that the EC had given only reluctant support to the imposition of sanctions on Iran following the 1979–80 hostage crisis. EC support for the peace initiatives in Nicaragua was also seen as unwarranted interference in events which were firmly in the American sphere of influence. In effect, Europe was gradually becoming a diplomatic and political force as well as an economic one. It represented an unwelcome challenge for USA governments faced with a decline in American hegemony, and with Japanese competition.

In summary, it can be argued that, by the mid-1980s, a strong conjuncture of forces had made necessary a fundamental reshaping of the Community. Despite American fears, it was the EC which was falling behind in the economic superpower conflict. It was clearly failing to fulfil its real potential, especially in relation to high-technology industries. The imperfections of the common market, combined with a lack of a coherent industrial policy, meant that Europe had been left 'with a population of sleeping industrial giants who were ill-equipped to meet the challenge of the 1970s and 1980s' (Geroski and Jacquemin 1985, 175). The EC was becoming a force at the international diplomatic level, but was considerably hamstrung by its institutional weaknesses and national rivalries. The Council of Ministers was barely functioning as a strategic decision-making body. It had become enmeshed in successive political conflicts over the budget and the details of policies such as the CAP. The second enlargement was considered likely to further exacerbate these difficulties and there was open speculation that the EC would split into inner and outer groups of full and partial members. This raised the prospect that the Community could become little more than an association centred on Germany. However, the future of the EC was to be far more complex than was suggested by these pessimistic forecasts. The mid-1980s were to be a turning point which witnessed three major shifts: the strengthening of technology policy, steps towards political union, and the 1992 Single Market programme. These are discussed in the remainder of this chapter.

Technology: Competition and Collaboration

One of the responses of the EC to the global challenge of the USA and Japan was to develop a positive technology policy. There was a concerted attempt in the 1980s to promote joint research and development initiatives in 'new technologies', especially relating to information technology, biotechnology and new materials. The idea of developing an EC technology policy 'has never been far below the surface for those seeking to create a united Europe. Indeed, it formed an explicit part of the programme of Jean Monnet's Action Committee for a United States of Europe' (Sharp 1989, 203). However, this was constantly thwarted by national and inter-company jealousies, especially over the role of 'national champions'. For example, an attempt to establish collaboration between Siemens, Bull and CII on computer research failed when it became known that the French government had negotiated a secret deal with the American company Honeywell, without informing the German government. In addition, West Germany generally favoured non-interventionist industrial policies, and was reluctant to see the Community become involved in technology programmes. With the exception of EURATOM, therefore, EC industrial policy was almost entirely restricted to eliminating barriers to trade and competition.

There was also a degree of complacency amongst EC companies and governments. The growing strength of Japan in R&D was not fully appreciated, at first. Yet in the mid-1970s Japan had launched its Very Large Scale Integration Programme, which brought together its main electronics companies with the aim of breaking into large-volume electronic-chip production (Sharp 1989). Similarly, there was a prevalent but false myth that the USA was falling behind in technological research (Strange 1987). This was not the case, for American research had strength in depth: a large home market, a large defence market, and large well-endowed universities. European illusions were rapidly disappearing by the late 1970s and the early 1980s, when there was a realization that the EC significantly lagged behind in new technologies. This did not apply to all sectors for EC companies were competitive in areas such as chemicals (FR Germany), pharmaceuticals (UK) and nuclear energy (France). However, the EC had fallen behind in key sectors such as consumer electronics, computers and telecommunications.

The structural weakness of the EC was starkly highlighted by the global trade statistics for high-technology goods. Whereas in 1988 Japan had a surplus of $8.6 billion and the USA of $1.3 billion, the EC had a trade deficit of $10 billion (Madelin 1988). The Commission of the European Communities also stressed the bleak long-term outlook for Europe:

> of the 37 technological sectors of the future that have been identified, 31 are dominated by the United States of America, 9 by Japan and only 2 by Europe; software and electronic switching (some sectors being dominated by two countries equally). In 1986, 4 out of 5 patent applications for new materials were filed by US or Japanese companies. (1988d, 8)

Why did the EC suffer from this technological lag? It was not because Europeans were less good at research. For example, in the period 1950–87 Europe produced 86 Nobel prize winners for science, compared to 115 in the USA and only 3 in Japan (Strange 1987). It was also not attributable to the level of expenditure on R&D. The EC, as a whole, did lag behind its rivals in terms of the proportion of GDP spent on R&D (figure 5.3a). At the level of individual states, even Germany was unable to compete with Japan in the key US market (figure 5.3b,c). However, the expenditure gap between the leading EC states and the two global giants was not especially significant. Similarly, the lag is not attributable to the level of government support. National governments in the EC12 planned to spend £320 billion on research in the period 1986–91,

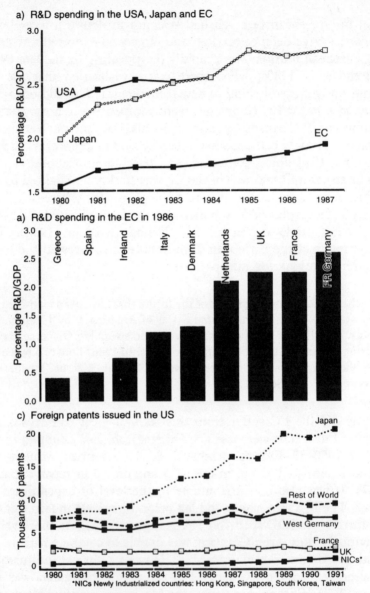

a) R&D spending in the USA, Japan and EC

a) R&D spending in the EC in 1986

c) Foreign patents issued in the US

*NICs Newly Industrialized countries: Hong Kong, Singapore, South Korea, Taiwan

Figure 5.3   R&D expenditure and patents in the EC, Japan and USA
Sources: (a) and (b) Commission of the European Communities (1988b);
(c) *Financial Times*, 9 June 1993

which exceeded the £230 billion support offered by the Japanese government although it was less than the planned spending of £700 billion by the American government (*Financial Times* 13 April 1988).

The root cause of EC weakness was its less effective application and commercialization of R&D. European R&D was fragmented along national lines and most EC companies were too small to compete in the global marketplace. Enormous resources and innovative ability are required to be able to bridge the gap between the laboratory and the marketplace quickly and effectively. EC companies lacked such resources and were weakened by their dependency on small national markets and by wasteful duplication of research.

By the 1980s attitudes in Europe were becoming more receptive to international collaboration, especially in view of the widely discussed potentials of the American 'Star Wars' initiative and Japan's fifth-generation computer R&D programme (Shearman 1986). In addition, the prospective deregulation of the giant AT&T company would soon allow it to compete outside the USA, while American anti-trust authorities were allowing IBM to enter the telecommunications sector. There was also a new Commissioner, Vicomte Davignon, in charge of the EC's industrial policies, and he was 'determined to transform its image from one of supporting lame ducks (or condoning such support from national governments) to one of actively helping to create a new industrial structure for Europe' (Sharp 1989, 207).

The principal outcome of these changing circumstances was the launching of the EC's ESPIRIT programme, in pilot form in 1982, and fully in 1984 with a budget of 1.5 billion ECU. ESPIRIT was 'to encourage transnational communications, common technical standards and collaborative approaches to IT through cross-border collaboration'. By concentrating on the pre-competitive stage of developing new technologies, it hoped to avoid conflict with its own competition rules and to persuade companies to put aside individual commercial jealousies. The main areas of research were advanced microelectronics, software technology, information processing, office systems and computer-integrated manufacturing. It proved to be a major success with, for example, Thomson (France), Plessey (UK) and Siemens (FR Germany) achieving a critical breakthrough in developing chips for a new generation of supercomputers. This encouraged the Commission to launch the second phase of ESPIRIT, with a budget of 1.6 billion ECU for 1988–91, followed by a third phase in 1990–4 with a budget of 1.35 billion ECU.

ESPIRIT was important in demonstrating that the EC did have a role to play in technology research. However, it was not the only EC programme; some other ventures were (see Sharp and Pavitt 1993, 138–9):

- *RACE* launched in 1985 to develop the technological base for an Integrated Broadband Communications Network for Europe. It had a budget of 500 million ECU in its second phase, 1987–92 and of 489 million ECU, 1992–4;
- *BRAIN*, a joint venture in neuro-computing;
- *BAP*, the Biotechnology Action Programme, with a budget of 75 million ECUs in 1986–90;
- *APOLLO*, for research on satellite transmissions;
- *JET*, the Joint European Torus for research on thermo-nuclear fusion;
- *JESSI*, the Joint European Submicron Silicon project;
- *BRITE*, the Basic Research in Industrial Technologies for Europe project to encourage the application of new technologies in traditional industries. In its first phase, 1985–8, there was a budget of 125 million ECU with 450 million ECU for 1989–92;
- *TELEMATICS*, for telematic systems in health, transport, etc;
- *BIOTECH*, for basic research and safety assessment in the biotechnology field;
- *FAST*, for Forecasting and Assessment in Science and Technology;
- *COMETT*, the Community Programme for Education and Training in Technologies, with a budget of 200 million ECU, 1990–4;

- *EURAM*, for European Research on Advanced Materials;
- *CUBE*, the Concentration Unit for Biotechnology;
- *SPRINT*, the Strategic Programme for Innovation and Technology Transfer in Europe, which aims to help small firms to access specialist services;
- *ESA*, the European Space Agency.

These are all essentially what Sharp and Pavitt (1993, 137) refer to as 'diffusion oriented' schemes. As a result of the success of the ESPIRIT programme, the Commission suggested in 1986 that all the Community's R&D projects should be grouped together in a single 'Framework' budget. A total budget of 5.6 billion ECUs over four years was eventually agreed for this. This represented a major increase in Community spending on R&D and, therefore, could be seen as a high point in the development of the EC's technology policies.

However, the EC's technology policies are flawed in several important respects. Firstly, the level of resourcing is still relatively small and only amounts to 2.5 per cent of the Community's total expenditure. This is only a tiny fraction of the spending by national governments on R&D (Harrop 1989, 98). Secondly, the programme has been criticized for failing to address the particular problems of small and medium-sized enterprises. This was only partly remedied by the establishment of the Seed Capital and Eurotech Capital projects to provide venture capital for such firms (Rothwell and Dodgson 1990). Thirdly, the initiative has also been weakened by national rivalries and, for example, the Framework budget was delayed for nine months and substantially reduced in size because of UK opposition. More generally, as Shearman writes, the programme is riddled with political compromises:

> European collaboration in computing and telecommunications at the governmental and EC level has taken the form of a political process shaped largely by considerations of political influence and prestige within the international arena and arising from different access to, and facility with, technological development. (1986, 161)

Despite these flaws, the EC's technology programmes have contributed to a restructuring of the new technology industries in Western Europe. Companies such as Siemens, Phillips and Thomson have been transformed from national champions to global leaders in several major sectors (see Williams 1987, chapter 5). While the EC's technology programme is too limited to account for this, Sharp (1989) argues that it has been important psychologically: it has created channels for co-operation, convergent expectations about the future amongst top-level decision-makers, positive experiences of co-operation, and pressure for further elimination of national barriers to trade. The last point is probably the most critical in terms of the long-term development of the EC. Establishing co-operation at the R&D stage generated pressure for maximizing the possible commercial applications. This meant increased demands for genuinely open markets and the adoption of Europe-wide standards, so that collectively developed technologies could be sold to pan-European markets. This added yet further pressure for the completion of the Single European Market. It was appropriate, therefore, that when the Single European Act was passed (see p. 116) – in effect providing the legal framework for the Single Market – it incorporated the aim of establishing a European Research and Technology Community. Science and technology policies have subsequently been brought together into the Framework programme. Yet Sharp and Pavitt (1993, 149) still conclude that EC policies have 'yet to adjust themselves to the realities of the 1990s'.

Completing the Single Market

Political rather than economic union provided the initial stimulus for the 1992 Single Market programme. Under Spinelli's guidance, the European Parliament had adopted the European Unity Treaty in 1984, as a commitment to European unity. This led the European Council at Fontainebleau to establish two committees to consider the issues of European union. The Dooge Committee examined institutional affairs and the Adonnio Committee looked at ways of creating a 'People's Europe', with a common coinage, stamps and other symbols of unity. However, by the 1985 Luxembourg intergovernmental conference, the agenda had shifted to economic union. There was agreement on a programme to complete the Single Market by 1 January 1993. The member states also agreed to sign the Single European Act; this amended the Treaty of Rome and introduced majority rather than the normal unanimous voting for most of the measures necessary to implement the Single Market programme.

In a sense, the 1992 programme was no more than a logical extension of the traditional, negative free-trade ethos of the Community. The partial removal of barriers to trade, and the common external tariff, had led to trade-creation and trade-diversion effects within the EC. For example, between 1958 and 1986 the proportion of the exports of EC6 members which was destined for other member states had doubled (Wijkman 1989). The first and second enlargements had also given a boost to trade integration within Western Europe. However, by the mid-1980s there were clear signs that economic integration was at a standstill, and that there were many non-tariff barriers within the EC.

By the 1980s attention had also switched to the need to improve the global competitiveness of the EC. Unfavourable comparisons were made between the fragmented national markets of the EC and the large single market in the USA. As one senior European executive noted wryly: 'If America really wants to do something to help Europe catch up, it should start by introducing different currencies in its 50 states and by imposing proper frontiers between them' (quoted in Madelin 1988, 9). Many European companies had become global players, but they were still at a disadvantage compared to American companies for their domestic (EC) market was fragmented into protected national segments. This was critical for high-technology goods. The large single USA market allowed American companies to introduce innovative products several years before this was commercially possible in the EC. 'By the time sufficient demand arose in Europe to support local production, new

Table 5.2   Macro-economic consequences of the Single Market programme
for the EC in the medium term

	Customs formalities	Public procure- ment	Financial services	Supply- side effects	Average value
GDP (% change)	0.4	0.5	1.5	2.1	4.5
Consumer prices (% change)	− 1.0	− 1.4	− 1.4	− 2.3	− 6.1
Employment change (thousands)	200	350	400	850	1,800
Budgetary balance (% point of GDP)	0.2	0.3	1.1	0.6	2.2
External balance of trade (% point of GDP)	0.2	0.1	0.3	0.4	1.0

Source: Cecchini (1988, 98)

European firms were already at a competitive disadvantage relative
to American subsidiaries' (Geroski and Jacquemin 1985, 194).

These economic arguments were encapsulated in an EC report
(summarized in Cecchini 1988) which sought to estimate the costs of
the 'non-single market', that is, of the continuance of national
fragmentation. There were estimates that the benefits of a once-and-
for-all removal of market barriers would be equivalent to an
economic stimulus of 200 billion ECU. It was argued that 'ever-
present competition will ensure the completion of a self-sustaining
virtuous circle' (Cecchini 1988, xix). Prices would fall, demand and
output would increase, as also would productivity and R&D. This
would produce a supply-side shock equivalent to 4.55 per cent of the
GDP of the EC. Table 5.2 details the individual effects of the
removal of customs formalities, elimination of public-procurement
national favouritism, the deregulation of financial services and
general supply-side effects. There would also be a price deflation
equivalent to 6.1 per cent, substantial savings on public sector
budgets and improvements in the EC's aggregate external trade
balance. In addition, there was the prospect of creating almost
two million new jobs, at a time when the EC had substantially

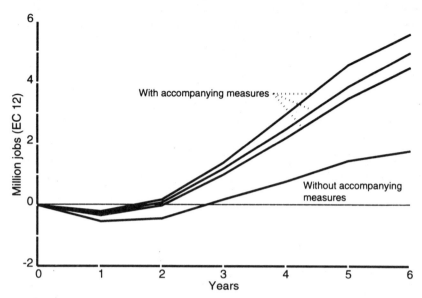

Figure 5.4   Employment prospects in the EC Single Market
Source: European Economy, no. 35

lower employment growth than either the USA or Japan (figure 4.1). If the productive resources released by these reforms were channelled into additional expansionary economic measures, the overall employment gains could exceed five or even six million new jobs (figure 5.4). The Cecchini Report was successful in capturing media and policy-making attention because it presented such a simple and positive message. Subsequently, economists have been less than flattering, with Baimbridge and Burkitt, for example, claiming that it 'possesses no theoretical back-up nor empirical content and is based upon implausible assumptions' (1991, 16). In short, the Report aggregated all the possible benefits but as the product of a long and implausible hypothetical chain of decision-making, investment decisions and so on. In addition, the removal of formal barriers was always likely to be partial, given compromises to protect national interests and, also, cultural differences. As Wise and Gibb (1992, 104) argue, 'whilst "1992" will create a mass market of 320 million people, it will not create a market for mass-produced goods'.

Putting aside such criticisms, the anticipated economic gains can be illustrated by reference to the car industry. It has been estimated that the Single Market programme could lead to a 4–5 per cent reduction in the costs of car production in the EC (Emerson et al. 1988). The 5.5 billion ECU of savings would result from a 2.6 billion ECU reduction in fixed costs, 0.9 billion ECU in variable costs and 2 billion ECU in marketing and distribution costs (Ludvigen Associates 1988). These cost reductions would be achieved via the elimination of technical, fiscal and physical barriers to trade, including:

- lack of a single EC-wide system of testing and approval;
- national variations in exhaust emission standards;
- unique national vehicle equipment requirements, such as yellow headlamp bulbs and side repeater flasher lights in some countries;
- variations in levels of taxes on car sales from 12 per cent in Luxembourg to 173 per cent in Denmark (see figure 5.5);
- different levels of annual licence fees or use taxes on cars;
- excessive state aid to national champions, such as that paid to the French company Renault in the late 1980s;
- border crossing costs, such as the administrative and delay costs of customs.

The cost reductions were important not only as an end in themselves, and as a positive benefit to consumers: they were also considered to be essential in improving the efficiency of EC car companies – in terms of productivity and R&D – so as to be able to compete with Japanese producers (Commission of the European Communities 1988d). An EC report has shown that Japanese companies have a number of critical advantages over EC producers. They have more beneficial links with component makers, have

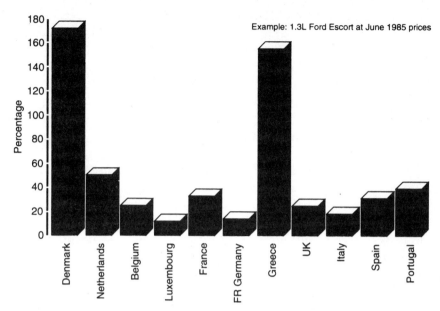

Figure 5.5   Tax-rates for new cars in the EC member states in 1985
Source: Commission of the European Communities (1988c)

higher labour productivity, higher quality production controls, better quality R&D, and more global sales and production strategies. The anticipated 5 per cent reduction in production and distribution costs was unlikely to modify significantly the competitive position of the EC in the face of such Japanese advantages. There was, however, a presumption that the Single Market would have a longer-term dynamic impact on the industry, leading to major productivity and R&D improvements. The critical question was whether EC producers had the capability to respond with the necessary dynamism.

The Single Market was also expected to eliminate inefficient practices in other industries. There were major and costly differences in national procedures for testing and approving pharmaceutical products which, taken together with state controls on marketing, led to enormous variations in final product prices. One product, Zyloric, cost 47 ECU in Ireland but only 5 ECU in Spain (Emerson et al. 1988). Creation of a Single Market would also lead to anticipated savings of 3 per cent of the costs of business services. In this instance, the major barriers were differences in legal systems, technical regulations and the recognition of professional qualifications. Finally, a comparison of domestic-appliance manufacturing in the USA and the EC is instructive. These markets were relatively similar in size but in the USA there were four producers, compared to 300 in the EC (Price 1988, 26). There was an assumption that the Single Market would lead to greater economies of scale and, hence, to lower costs and increased efficiency and innovation. While it is true that critical mass was necessary for R&D leadership in some industries, this did not apply to all sectors. There were some industries, such a clothing manufacture, in which the flexibility of small firms offered important competitive advantages. Indeed, the hypothesized relationships between size, efficiency and innovation are still largely untested (Mason and Harrison 1990). However, before proceeding to a fuller assessment of such implications, it is necessary to outline the Single Market programme in more detail.

The 1986 Single European Act defined the internal market as 'an area without internal frontiers in which the free movement of goods, persons, services and capital is ensured' in accordance with the Treaty of Rome. The process of creating a Single Market had begun in the 1960s with the elimination of most tariff barriers. However, there still remained in place a large number of additional barriers to trade, including state procurement, technical standards, taxation differences, and customs and other frontier formalities. A 1985 White Paper produced by Lord Cockfield for the Commission identified 279 measures which were necessary to remove such barriers in order to create a genuine Single Market. These can be grouped into physical, fiscal and technical barriers.

- *Physical barriers.* The elimination of all frontier controls on the movement of goods, people and capital such as customs duties, passports and health checks. These apparently simple measures were controversial as they involve a loss of national controls

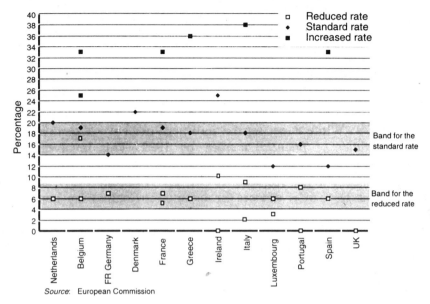

Figure 5.6   VAT rates in the EC12, and Commission proposals for
new bands
Source: Bos and Nelson (1988)

over the potential movement of animal diseases, illegal migrants
and terrorists.

- *Fiscal barriers.*   The harmonization of value added taxes (VAT)
  and other purchase taxes in the EC. There was a considerable
  spread in the rates of VAT and in the products to which these
  were applied (figure 5.6). Therefore, the aim was to agree two
  bands of VAT rather than specific rates: a standard rate of 14–20
  per cent and a reduced rate of 5–9 per cent. Even this
  compromise caused considerable dificulties for some member
  states. Denmark faced the greatest adjustment because it had a
  single high rate of VAT at 22 per cent, had no VAT on several
  essential products and had high excise duties, such as that on
  cars referred to earlier (Guieu and Bonnet 1987). Another
  contentious issue was whether to collect VAT at frontiers or in
  the receiving country (Bos and Nelson 1988). The October 1989

Council of Ministers meeting chose the latter solution which, although administratively cumbersome, was considered to be more secure against fraud.
*   *Technical barriers.* These include a diverse set of barriers to trade relating to standards, regulations and subsidies (discussed below).

The 1985 White Paper set the ambitious objective of completing *the integration of financial markets* by 1992. This involved the elimination of all restrictions on long-term and short-term capital movements. There was also to be full freedom for financial intermediaries to offer services throughout the Community, and this required the right to establishment and the removal of restrictions on cross-border transactions. By 1988 agreement had been reached to liberalize capital movements by 1990, with Spain, Portugal, Greece and Ireland being given extensions until 1992. Arguably, this was the single most important measure in the entire package. It made essential the harmonization of taxes and significantly increased the pressure for a common monetary policy. The harmonization of financial markets has also been problematic. With banking and securities dealing, the approach was to agree minimal rules and standards but to place regulation in the hands of national bodies. For non-life insurance services, a 1988 directive agreed liberalization of cross-border sales by the end of 1990. However, it has been far more difficult to agree rules for harmonizing the sales of life assurance. But, in general, there has been a remarkable liberalization of financial services in the 1980s and 1990s, with most of the Single Market objectives being achieved ahead of schedule. However, as Leyshan and Thrift (1992) remind us, many of these changes were due to global pressures in the sector which, in some instances, made the 1992 programme seem outdated.

Also covered was *technical harmonization* so that common standards of design, safety and other specifications apply throughout the EC. This means that, for example, goods would only have to be tested and approved once in one EC country, rather than be subject to separate assessments and testing in each member state. According to the principle of 'mutual recognition' (Pinder 1989), it would be possible to sell throughout the EC any object legally manufactured and marketed in one member state. Divergent technical standards are especially important in the manufacture of cars, electrical engineering, mechanical engineering, pharmaceuticals, non-metallic mineral products and transport equipment (Cecchini 1988).

An estimated 15 per cent of the GDP of the EC was composed of goods and services which were covered by *public procurement* (*Financial Times* 27 June 1988). Public sector procurement was usually guided by overt or latent national favouritism, which was reflected in the fact that 90 per cent of all public-sector equipment purchases in the EC are from national suppliers. This constituted a major obstacle to free trade, especially in the supply of energy, water transport and telecommunications.

There were still institutional barriers to the *free movement of labour*, particularly the recognition of professional qualifications. It was proposed that there should be mutual recognition of virtually all such qualifications. However, this is a difficult process; for example, it took seventeen years to reach agreement on architects' qualifications.

As of 1993 quotas on road vehicles operating in other member states were eliminated, hence opening up the possibility of a single market in road *transport*. There was also agreement that there would be freedom to provide shipping services to, from and between member states as of 1993. Air transport remains dominated by private and public cartels and has proved less easy to liberalize. Transport therefore remains one of the weak links in the 1992 programme; little has been achieved in the way of eliminating barriers and restrictions. This was compounded by the lack of an effective Common Transport Policy which could contribute positively to greater economic integration as well as to the aims of environmental and social programmes (Whitelegg 1988).

State subsidies provide another barrier to trade; these varied considerably in importance amongst the member states, ranging from 5 per cent of domestic income in France to 11 per cent in Ireland (Lintner 1989b).

MCAs constituted barriers to *agricultural* trade. There was agreement not to introduce any new MCAs and to phase out existing ones during the 1990s.

There were moves to open up the national markets for *telecommunications*, especially for the provision of value-added services rather than basic telephone services.

The Commission aims to create a '*European broadcasting area*' by removing legal and technical barriers and by agreeing rules and standards for advertising and programme content.

One of the aims of the Single Market programme was to shock the Community out of its complacency. Delors expressed it thus: 'In the world race against the clock, which the countries of Europe have to win to survive, what was needed was a common objective to enable us to look beyond the everyday difficulties and pool our strengths and energies' (Commission of the European Communities 1988c, 5). In strategic terms these objectives have been fulfilled, but implementation of the Single Market programme has been an uneven process. Approximately 60 per cent of the 300 measures had been passed by the Council of Ministers by March 1990. Inevitably, these tended to involve the less contentious and simpler directives, especially those that could be made on the basis of majority voting, as specified under the Single European Act. Thereafter, the process slowed as more controversial matters were addressed, such as taxation, which required unanimous voting. By June 1991 only 73 per cent of measures had been approved. This led to a flurry of hasty decision-making in the final eighteen months to ensure that the Single Market measures were agreed by the end of 1992; in the event eighteen remained unresolved by the deadline. This inevitably meant there would be a number of compromises, so that some measures – such as that on VAT rates – embodied little more than the spirit and none of the required detail for a genuinely open market. There were also long transitional periods for the adoption of some measures; for example, Germany does not have to implement fully the directive on the taxation of multinational branch plants until 1996. Furthermore, many national legislatures lagged behind in transposing Council decisions into national laws; this was especially the case with Italy which in mid-1992 had transposed fewer than half the measures accepted by the Council (Ahlstrom 1991). As a result, the Single Market which came into existence on 1 January 1993 was, perhaps inevitably, both a broader and a shallower economic space than had originally been conceived of in the 1985 White Paper. Even so, it had much wider social, political and economic implications, and it was for this reason that it had become known as the 'Russian dolls' strategy.

'Russian Dolls': the Single Market as a Catalyst

In some ways the Single Market programme could be seen as a 'tidying-up' operation designed to fulfil some of the original objectives of the Treaty of Rome. The White Paper represents:

> a plan for a large scale onslaught on the 'omissions' in the *acquis* with respect to Treaty obligations crucial to market integration, and a simultaneous endeavour to make progress on issues where the Treaty provides a permissive legal basis but no obligation. (Pelkmans and Robson 1987, 193)

There is no doubt that, collectively, the 279 measures will boost economic integration. Whether they will achieve the predicted 4.55 per cent increase in GDP and a two million increase in employment is at best uncertain and, indeed, these forecasts have been criticized. There can be even less confidence that the dynamic gains will materialize. As Cecchini has written: 'the gains forecast for the European economy will not appear as if by divine intervention. Realising the potential that is on offer presupposes a robustly positive response to the supply-side opportunity by business and government' (1988, 72). However, the scale of both the immediate and the dynamic effects is, in a sense, less important than a number of indirect economic and political consequences emanating from the 1992 programme.

Some of the most important consequences relate to the *political implications of the 1986 Single European Act.* This has started to unravel the stranglehold imposed on decision-making by the Luxembourg compromise. It commits the member states to majority voting on most of the measures required for the 1992 programme; the areas excluded relate mainly to taxation and to health matters. The system is one of qualified majority voting. A total of 76 votes are distributed amongst the member states and 54 votes are required for a qualified majority. Votes are distributed in relation to population size and it would require at least two large states (such as the UK and FR Germany) and one smaller state (for example, Denmark) to muster a blocking minority. Majority voting was, of course, essential for the very success of the 1992 programme:

> Without the Single European Act or the recognition that twelve very different countries could undertake serious and enforceable political and legal obligations, the Cockfield White Paper of June 1985, *Completing the Internal Market,* would have been just another statement of forgotten desiderata. (Wallace 1988, 177)

The longer-term significance of this is that member states have conceded an important element of political sovereignty, and this is likely to affect the entire future pattern of decision-making in the Community. The Single European Act also changed the balance of power between the Commission, the Council of Ministers and the European Parliament in other important ways (see pp. 201–8). These political implications were no accidents, but were part of what became known as the Commission's 'Russian doll' strategy. The Single European Act included vague but significant commitments to new 'competences' for the EC. As, will be seen in the following chapter, 'Hardly had the ink dried on the Single Act when the Commission and its leaders were prodding Europe towards newer and wider aspirations' (Ross 1991, 59).

The 1992 programme represented a challenge to the *economic sovereignty of the state* as well as to its political power. Member states conceded power over several key areas of economic policy. This was a fundamental challenge to the models of capitalism which had

Table 5.3   State aids to manufacturing in the EC, 1981–6
(excluding steel and shipbuilding)

	% of value added	ECU per worker
Italy	15.8	5,951
Greece	13.9	–
Ireland	12.3	3,741
Belgium	4.5	1,373
Netherlands	4.1	1,419
France	3.6	1,223
Luxembourg	3.5	1,079
UK	2.9	757
FR Germany	2.9	940
Denmark	1.7	609
EC average	5.5	1,774

Source: Quoted in *Financial Times*, 20 November 1989

evolved in most EC states since 1945, that is, of free enterprise mixed with state macro-economic regulation and (often-subsidized) state-owned industries. The challenge to national economic sovereignty was formidable and included such critical areas as taxation, customs and excise, and the use of MCAs to support farmers or consumers. However, it was most clearly evident in the EC challenges to state subsidies and to public procurement policies, both of which are contrary to the competitive and single-market ethos of the 1992 programme.

State aids to manufacturing were and are widespread throughout the EC, and represent subsidies to both publicly and privately owned companies. Their extent varied between member states (table 5.3): at one extreme were Italy, Greece and Ireland, where state aid exceeded 12 per cent of value added, compared to less than 3 per cent in Denmark, the UK and FR Germany. The Commission was concerned that, with the removal of most other forms of non-tariff barriers to trade, governments might have been tempted to increase their use of state subsidies as a form of protectionism. This was particularly likely where national champions or declining industries concentrated in regions with high unemployment were concerned. The Commission has, therefore, taken an increasingly tough line with member-state governments. Some of the more important test cases of the late 1980s have been French aid to the Renault car company, Italian refusals to close the Bagnioli steel works as part of an agreed restructuring of the steel industry, and the UK government's subsidized sale of the Rover car company to British Aerospace.

Public procurement was another sensitive area of protectionism, not least because an estimated 15 per cent of the Community's GDP was accounted for by this. A number of manufacturing industries were involved, including boilermaking, turbine generators, loco-motives, mainframe computers, telephone exchanges and telephone handsets (table 5.4). With the exception of computing, there was relatively little intra-EC trade in these industries. For economic and/or strategic reasons a large part of their output was influenced by public procurement procedures and national favouritism. In consequence, there were far larger numbers of producers in most of these sectors in the EC than there were in the American market. It was argued that the liberalization of trade, followed by increased concentration and greater economies of scale, could have yielded cost reductions of 5–20 per cent in these industries.

As part of the 1992 programme, member states agreed to increase international advertising of and tendering for larger contracts involving public procurement. The Commission has also tightened up its existing, but largely ineffectual, rules on competition and public procurement and has the power to take offenders to the European Court of Justice. However, the effectiveness of these measures depends on aggrieved parties being willing to lodge formal complaints. Most are unwilling to do so, given their dependence on the member states for future contracts. Complaints to the Commission are rare; one exception is that by Bouygues, the French construction company, that a Danish bridge-building contract had insisted on the use of Danish labour and materials. Even though informal protectionism will continue, greater competition is likely to lead to restructuring in these sectors. It is partly in response to such

Table 5.4 Key industries affected by public procurement policies

	Community market (ECU)	Current capacity utilization	Intra-EC trade (per cent)	Number of EC producers	Number of US producers	Economies of scale (per cent)[1]
Boilermaking	2 bn	20	very little	12	6	20
Turbine generators	2 bn	60	very little	10	2	12
Locomotives	100 m	50–80	very little	16	2	20
Mainframe computers	10 bn	80	30–100[2]	5	9	5
Telephone exchanges	7 bn	70	15–45	11	4	20
Telephone handsets	5 bn	90	very little	12	17	–

[1] Unit cost reduction resulting from a doubling of output
[2] Percentage of total demand
Source: EC Commission; *Financial Times*, 27 June 1988.

pressures that, for example, major companies in the turbine generator sector have been investigating possible mergers and new forms of co-operation. Amongst these is the merger between Brown Boveri of Switzerland and ASEA of Sweden, two companies based outside the EC but with major interests in EC markets.

Even if the Community is successful in weakening protectionism in these manufacturing industries, there are even greater challenges in the real heartland of public procurement: energy, transport, telecommunications and water. These are all highly sensitive areas of national economic sovereignty. Yet, in February 1990, agreement was reached on opening up all four sectors, except oil, gas and coal exploration. In practice, however, the degree of opening will be limited because there are other obstacles to trade and competition. These include in-house procurement by large companies, the lack of European standards for railway track gauges, and divergent national telecommunications standards. In telecommunications only one-third of the entire market – advanced telecomms and basic data communication – have been opened up to competition. While this demonstrates that the Single Market will not reach into all areas of production and distribution, it does also highlight the loss of national economic sovereignty in even the most sensitive of areas.

Although many of the proposed directives would not take effect until the 1990s, their influence on corporate *merger and acquisition activity* was already apparent in the late 1980s. The 1992 programme has had an important psychological impact on a process of corporate restructuring which was already under way. Companies have had to look beyond their domestic markets and have had to develop longer-term strategies. This concentrated their minds not so much on scale *per se*, but on the question of market share (*Financial Times* 5 October 1987). The outcome has been a spate of mergers, acquisitions and joint-ventures. This is a more efficient way to build up market-share than establishing completely new foreign networks. Although there is no clear evidence that such mergers and acquisitions necessarily lead to improvements in efficiency (Cowling 1980), they are viewed as essential for long-term development and survival strategies.

Table 5.5   Cross-border acquisitions in Europe in 1989

	ECU million	Total no. of deals
Target nations		
UK	20,831.8	237.8
FR Germany	5,710.3	215.9
France	5,366.0	191.4
Italy	4,121.9	104.1
Spain	2,689.4	128.4
Netherlands	1,883.3	98.5
Belgium	1,285.6	61.9
Sweden	762.1	34.9
Denmark	543.8	34.5
Acquiring nations		
US	13,803.2	185.1
France	9,674.4	167.5
FR Germany	6,647.0	128.5
UK	5,512.0	281.6
Italy	1,681.4	52.0
Japan	1,481.6	54.6
Sweden	1,381.6	120.7
Belgium	1,016.3	27.9
Switzerland	926.4	82.9

These tables include acquisitions made in European nations by US and Japanese companies.
Source: Translink's *European Deal Review*, quoted in *Financial Times*, 5 February 1990.

The cross-border acquisitions reveal a distinctive geographical pattern. There were more than 100 cross-border acquisitions within the EC in 1989 alone, and an even larger number in Western Europe as a whole. The main targets were the UK, France and the FR Germany (table 5.5). In part, this reflected the relative weights of these economies. However, the fact that the value of acquisitions in the UK was five times larger than that in FR Germany also reflected the liberal company take-over rules in the former. Most of the purchasers came from the same three European countries, together

Table 5.6   European merger and acquisition activity in 1988

Type of transaction*	$ billion	No. of deals
Domestic in Europe		
(within one country	59.6	100
Transatlantic	52.6	101
Cross-border with Europe	13.2	24
Europe–Asia	8.8	15
Total	134.2	240
US domestic	202.5	323

* Only transactions over $100 million
Source: First Boston/CSFB, quoted in Financial Times, 11 January 1989.

with the USA. Indeed, in value terms, the USA accounted for more cross-border acquisitions within Europe than did the UK and FR Germany together. Japan and Switzerland also featured prominently in this league table, thereby underlining the wider global significance of the 1992 programme. The acquisitions included both manufacturing and service companies, with the largest acquisition in 1989 being the purchase by Victoire (France) of 42 per cent of the shares of the German insurance company Colonia.

There is, however, a danger of seeing mergers and acquisitions as a uni-causal process driven by the 1992 programme. This is far from the truth, as is evident in an analysis of the global mergers and acquisitions of European companies (table 5.6). In terms of value, the largest proportion of mergers and acquisitions in 1988 involved two or more companies within individual European countries. This was closely followed by transatlantic deals which, in value terms, were four times larger than cross-border activities within Europe. Even more striking is the fact that all mergers and acquisitions involving European companies were equal in value to only two-thirds of the activities within the USA economy. Mergers and acquisitions are a regular part of the global strategies of large corporations and the importance of 1992 has been to increase the role of Europe in these. Thus the ASEA–Brown Boveri merger, already referred to, was not principally inspired by the 1992 programme. Instead, its root cause was considerable over-capacity in the world's electrical engineering industry and the need to keep potential partners out of the ownership of major rivals. As *The Financial Times* (31 May 1988) commented on this and other major acquisitions:

> for many European manufacturers 1992 is mainly a state of mind which is accelerating their reaction to a much more fundamental trend: the opening of markets around the world to the power of global scale (in research, development and manufacturing), and geographic scope (in distribution and the ability to cross-subsidise from market-to-market).

In other words, 1992 was not about European competitiveness but about the global competitiveness of European companies. This, of course, was the link with the EC's technology programme. If programmes such as ESPIRIT, RACE and BRITE are to succeed commercially, then European companies have to be globally competitive. Mergers and acquisitions are one element in this. Another element is the need to harmonize company taxation and legislation across the EC so as to make it easier for genuine pan-European companies to emerge and operate. This is underlined by the bitter experiences of several major EC mergers in the early 1980s. For example, that between Agfa and Gevaert failed precisely because of such difficulties.

The Single Market has also had a catalytic effect on *inward investment* into the EC. Both the Japanese and the Americans were suspicious that the 1992 programme was little more than a recipe for 'Fortress Europe', which was seen as 'a giant industrial version of the Common Agricultural Policy' (Owens and Dynes 1989, 185). The USA was especially concerned about the EC demands for reciprocity: that EC companies should have the same access to markets in other countries as the MNCs of those countries enjoyed in the Community. This particular row was defused when, in 1989, the EC reinterpreted reciprocity as a 'broad national treatment approach'. EC companies were to have the same rights as national companies in foreign markets. Japan's fears ran deeper because it had been subject to increasing EC protectionism, such as anti-dumping directives, throughout the 1980s. Many member states had also imposed bilateral import quotas on Japanese products. While the UK had no formal import quota (but did have voluntary export restraints), one member state operated 41 such quotas. After 1992, bilateral import quotas would no longer be possible in the EC, but there was a real fear in Japan that these would be replaced by Community-wide restrictions on Japanese imports.

Partly because of their fears of protectionism, but also because of the perceived advantages of the Single Market of 320 million consumers, it was expected that inward investment to the EC would increase. In fact, American fears were ill-founded because of their already considerable presence within the Community. Indeed, the sales of the affiliates of American multinationals located within the EC were six times larger than direct US exports to the EC in the late 1980s (*Financial Times*, 8 May 1989). American companies have also developed more genuinely pan-European production and distribution strategies than have European companies. They were, therefore, considered to have the potential to be major beneficiaries of the 1992 programme.

Japanese companies, however, were in a weaker position. Their total cumulative investment in Western Europe was only $11 billion in 1985, compared to $107 billion by American companies. Unlike American companies, foreign investment was seen as supporting exports rather than as being a substitute for these. However, by the mid-1980s growing protectionism, including the perceived threat of 'Fortress Europe', led to a reassessment of Japanese strategies. In 1988–9 alone, the level of Japanese investment in most major EC states was equivalent to more than one-third of their cumulative investment in the whole of the period 1951–88 (*Financial Times*, 28 June 1989). This was particularly strong in the car industry and led to a wave of investments in car production capacity in the EC by

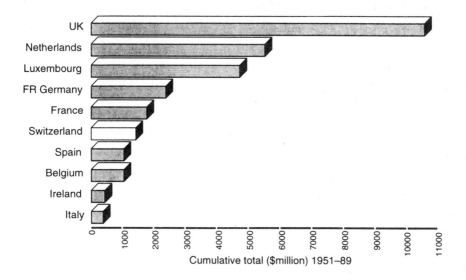

Figure 5.7(a)   Total Japanese investment in selected European countries,
1951–89
Source: Financial Times, 28 June 1989

companies such as Toyota, Nissan and Honda. Most of the investment was directed to establishing transplant factories in the UK, which continues to be the most attractive location for inward investment from Japan (figure 5.7a). The 1992 programme has, therefore, contributed to an important geographical reorganization of production in the EC. One of the implications of this is that the UK is becoming a net exporter of cars instead of a net importer.

Not surprisingly this has led to the accusation that the UK, in particular, has become a 'Trojan horse' whereby the transplants of Japanese and other foreign investors can exploit the single market. This led to European demands for protectionism, which largely centred on a debate about local-content rules. The EC was faced with defining the percentage of product content which must be sourced within the EC, if particular items were to be regarded as EC

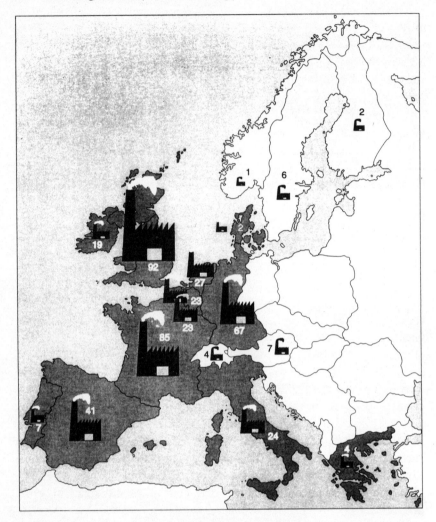

Figure 5.7(b)   Number of Japanese manufacturing plants in January 1989
Source: Financial Times, 28 June 1989

products. However, the debate was complicated by the geography of consumption as well as by the geography of production. Those countries without indigeneous car producers had no vested interests in keeping out the Japanese and, indeed, in many cases already had substantial Japanese imports. Thus Japanese producers accounted for 38 per cent of the Irish market compared to less than 2 per cent in Italy and Spain. Five EC member states had voluntary car export-restraint agreements with Japan, but these could not be sustained in the Single Market. The eventual agreement between Japan and the EC, in June 1991, froze imports at their current level until 1999, when all restrictions were to be removed (Dicken 1992). Cars produced in Japanese transplants, with at least 60 per cent local (that is, EC) content, were allowed open market access.

There is again a need to look at the global context of inward investment in order to place the effects of the 1992 programme in perspective. First of all, the EC, even in the late 1980s, still received only a small proportion of all Japanese investment; only 16 per cent in 1988 compared to the 42 per cent destined for the USA. Secondly, the EC was as much an exporter as an importer of capital. In 1986, for example, net inward investment from other EC states was greater than that from non-EC countries in the cases of France, UK, FR Germany and the Netherlands (table 5.7a). Even more significantly, all four states exported far more capital to non-EC countries (mainly the USA and EFTA) than they received from these (de Smidt 1992). The 1992 programme is contributing to the reorganization of the geography of international investment but it is only one element in a much larger global picture. This is confirmed by the sharp reduction in inward investment between 1989 and 1992 in the face of global recession, despite the approaching deadline for completion of the Single Market (table 5.7b).

With the virtual completion of the internal market in hand, attention switched to the question of economic federalism, whereby some key areas of economic policy and a common currency would be administered by a central EC body. In particular, this has resulted in growing pressures for *monetary union* in the EC, which constitutes another of Delors's 'Russian dolls'. The link between monetary union and the Single Market is best explained by reference to what has been described as an inconsistent quartet: free trade, capital mobility, exchange rate fixity and national monetary sovereignty

Table 5.7

a)   Foreign investment in Europe in 1986

| | Million ECUs | | | |
| | Net inward | | Net outward | |
	EC	Non-EC	EC	Non-EC
West Germany	1,155	−111	3,685	5,853
France*	1,428	1,390	1,802	3,534
UK	3,482	2,591	3,681	13,654
Netherlands	1,562	856	3,010	1,453

* Excluding reinvested profits

b)   Direct inward investment, 1989–92 (£ million)

	1989	1991	1992 (Jan–June)
France	9,552	11,109	7,660
Germany	6,997	2,200	1,573
Italy	2,529	2,542	1,727
Netherlands	6,772	3,700	2,977
Spain	8,433	11,100	3,775
UK	28,043	21,100	9,694

Source: European Commission

(reviewed in Padoa-Schioppa 1988). The Single Market was overtly designed to assure the first of these, free trade. Capital mobility has followed in its wake and the Council of Ministers agreed that there would be full liberalization of capital movements in most states by 1990, and in all member states by 1993. This increased the pressure on exchange rates, to counter which it became necessary to strengthen the co-ordination of national economic policies (Micossi 1988).

Most member states, with the belated addition of the UK, effectively adopted exchange rate fixity via their membership of the Exchange Rate Mechanism of the EMS. This was due to more than just political good will. If exchange rates were allowed to fluctuate then this would be a new way of fragmenting the Single Market just as the old barriers were being swept away. It would cause uncertainty in international trade and about the terms of access to the EC market – which provide the essential basis for the expected strengthening of competition, investment and R&D. In other words, there was a strong argument that monetary union was as essential to the success of the long-term objectives of the 1992 programme as was the removal of non-tariff trade barriers.

This general argument was accepted by all the member states, including the UK, at the 1988 Hanover European Council when it was agreed to speed up progress towards monetary union in the light of the 1992 programme. The Delors Committee reported in April 1989 and advocated a three-stage transition to full monetary union: that all EC currencies should belong to the EMS and the ERM; that there should be increased collective decision-making; and that there should be fixed exchange rates, preferably with a single European currency. While the economic pressures for full monetary union are considerable, there were also strong opposing political pressures. Macro-economic management was an essential election-winning instrument for EC governments, so some of the political parties in some member states were reluctant to relinquish their controls over this. As ever, the 1992 programme was as much about politics as it was about economics. This was probably the most potent of all the 'Russian dolls' and it was to provide one of the great dramas surrounding the Maastricht agreement (chapter Six).

Finally, we return to one of the fundamental objectives of the Single Market: will the 1992 programme lead to a closing of the *technology gap* between the EC and its global rivals? Firstly, it should be re-emphasized that the EC does not lag in all advanced technology industries. There are sectors in which the EC is at the technological forefront, such as aerospace, defence electronics, chemicals, pharmaceuticals and nuclear energy. However, the EC lags in many other sectors, and especially in electronics and microelectronics. The Single Market will create improved commercial conditions for companies operating in Europe. However, there is no guarantee that EC companies rather than Japanese or American ones will benefit. There is also no guarantee that Single Market conditions will not lead to greater concentration (related to hoped for economies of scale), and hence reduce the degree of competition in the EC. Ultimately, however, it is likely that the success of the Single Market programme will depend as much on global economic relationships as it will on economic and political developments within the EC.

SIX

Maastricht and Renewed Uncertainty: the 1990s

The 1980s began with deep pessimism about the future of the EC, but the decade closed with the Single Market programme being implemented and with renewed confidence about Europe's global role. In effect, there had been a significant deepening of the economic space of the Community via increased integration of trade, as well as some deepening of the political space via the Single European Act. It had long been the Commission's view that this would generate new creative tensions with the Community. A Europe-wide economic framework was being established in some areas of trade and regulation, which was bound to generate demands for complementary economic reforms, especially in terms of monetary union. There were also likely to be pressures for the Community to increase its competence in areas such as environmental standards, to ensure that there was an effective single economic space. But of even greater longer-term significance was the imbalance being created between economic and political spaces. In effect, an international economic space, with multinational rules, was still being regulated by intergovernmentalism based on the powers of individual states, combined with an increasingly powerful but essentially non-democratic Commission. There were also tensions between the EC's economic and social spaces. While the new economic space offered considerable advantages to capital – or at least to some sections of international capital – there was a welfare deficit in terms of the rights of workers and the needs of vulnerable and disadvantaged social groups. In other words, the creation of a single economic space generated pressures for the establishment of a single social space. The major question for the 1990s, therefore, was whether economic integration would be deepened and be accompanied by greater social and political integration. However, there was also to be another challenge for the Community in terms of widening the single economic space so as to include non-member states from the EFTA group.

The European Economic Area

One of the immediate effects of the Single Market programme was that it caused the EFTA countries to reassess their relationship with the Community. This was anyway a fluid relationship which had been subject to revision ever since the two major Western European international institutions had been created in the late 1950s. The 1970s and the 1980s enlargements of the EC had both transferred members from EFTA to the Community, leading to the need to reshape trade and institutional ties. The increasing integration of trade and the pivotal role of Germany in the European economy had also tied the two groups more closely together. By the 1980s two of the three most important trading partners of all the EFTA countries were to be found in the EC (Wijkman 1989). The need to review this relationship became more urgent for the EFTA countries with the approach of January 1993, and the realization that they would have to adopt the rules of the Single Market. There was also the negative impulse stemming from the fear of being locked out of 'Fortress Europe', in terms of both trade and foreign direct investment.

At first, the realignment between the two groups was sought on the basis of bilateral negotiations with the individual EFTA countries, but these proved to be problematic. In a characteristically bold move, the President of the Commission, Delors, then proposed that there should be direct negotiations between the two institutional blocks so as to create a new economic space. This was to be a half-way house between full membership and a treaty of association. Negotiations commenced on this basis in June 1990. At first they made rapid progress, but then became ensnared by the troublesome issues of fishing, road transport, and the payment of compensation to the poorer EC member states in return for the increased competition they were likely to face. Nevertheless, an agreement was signed in May 1992 establishing the European Economic Area, which was to come into effect in the following year. In the event, the Swiss electorate rejected the agreement in a referendum, so that when it finally came into existence in the second half of 1993, it applied only to Liechtenstein, Sweden, Norway, Iceland, Finland and Austria.

According to the joint press release issued in May 1992, the European Economic Area is:

> intended to give fresh impetus to the privileged relationship between the European Community, its Member States and the EFTA States, which is based on their proximity, the importance of their economic relations, their common values of democracy and a market economy and their common European identity.

The emphasis on values, democracy and proximity may seem puzzling, at first sight, given that this was essentially an economic agreement. However, it has to be seen in the context of the pressure on the EC for membership from Eastern and Central Europe, and from Turkey, Cyprus and Malta. It was, therefore, a statement of the conditionality of access to the economic space – let alone the political space (full membership) – of the EC. The advantages of the Single Market would not be available to all who knocked at the Community's door.

The economic space was to be anchored on the four basic freedoms enshrined in the Treaty of Rome – the mobility of goods, services, people and capital – although there were also provisions covering fishing, transport, cohesion funds, and acceptance of other EC agreements and rules. These are summarized below:

- *Freedom of movement of goods.* Customs duties on industrial products had mostly been eliminated as early as 1972, but it was now proposed to remove non-tariff barriers, such as state procurement and monopolies, and differential national technical and legal standards.

- *Freedom of movement of persons.* Rights of residence, of movement and to work were extended across the European Economic space. This was to include non-discrimination in social security entitlements, as well as the mutual recognition of qualifications. Switzerland was given a five-year transition period in view of the revolutionary change this would entail to its, traditionally, highly restrictive immigration policies.

- *Freedom of movement of services.*   There was to be freedom of establishment for banks and insurance companies, as well as acceptance by EFTA of EC regulations on such matters as single licensing and home-country controls over the European activities of financial companies. Some countries were permitted transitional arrangements for phasing in the opening up of these sectors to foreign takeovers.

- *Freedom of movement of capital.*   All barriers to capital movements were to be removed, with minor exceptions covering the real estate and fishing sectors.

- *Fishing.* Trade was to be liberalized over a period of transition. As a concession to the hard-pressed EC fishing industry, Norway also agreed to increase the share of its North Sea quota allocated to Community fishing fleets; this however, represented only a minimal increased from 2.14 per cent to 2.9 per cent.

- *Transport.*   This was one of the most contentious areas, as securing improved transit rights through Austria and Switzerland was vital for the functioning of the economic space of the Single Market. In the end, Austria offered a substantial increase in licenses to Greece but few other concessions. Switzerland also offered only minor increases in road transit rights, insisting instead on joint road–rail initiatives so as to ease environmental pressures in the Alpine passes.

- *Wider co-operation.*   The EFTA countries accepted EC rules on competition, the environment and several other areas, while also securing access to Community programmes for education, tourism and research and development.

- *Cohesion funds.* In order to compensate the poorer members of the EC for increased competition, the EFTA countries agreed to pay 500 million ECU in grants and 1500 million ECU in soft loans to Portugal, Greece, some regions in Spain, and the island of Ireland (the latter was a device to avoid being seen to offer assistance to the UK as such).
- *Institutional changes.* An EFTA surveillance authority was established to oversee the arrangements, together with a joint court, meetings of ministers and a joint parliamentary group.

The implications of the European economic space were expected to be greater for the EFTA countries than for the EC, given the relative economic weights of the two groups. The former anticipated significant gains in trade and competitiveness within the Single Market. However, some of the more protected EFTA markets, as in financial services, were likely to experience considerable competition from powerful EC multinationals.

Athough the EFTA countries anticipated the most immediate economic benefits from the new arrangement, it also offered advantages to the Community. Firstly, it would increase the economic space of the Single Market to encompass eighteen countries (nineteen if Switzerland had joined), with a population of 380 million and accounting for 43 per cent of world international trade. The EFTA countries already accounted for 26 per cent of the Community' s extra-EC trade (Church 1991). Secondly, it influenced the EFTA countries' potential future full membership of the EC in three ways: deflecting, enabling and reinforcing. The new arrangement provided a means of deflecting the growing demand for membership from the EFTA countries at a time when the EC was absorbed with internal matters such as economic and monetary union. At the same time, it enabled potential future membership by the EFTA countries by securing agreement on many of the critical trade issues involved in enlargement. It also reinforced the likelihood of further enlargement for the European Economic Area that involved

economic obligations for the EFTA countries without the political
rights that accompanied membership. In this respect, therefore, the
European economic space represented yet another 'Russian doll' in
the process of European integration.

Maastricht: the Russian Doll Strategy Revealed

The 1986 Single European Act was 'an unstable compromise'
(Newman 1993, 13). While it was hailed, at the time, as an advance
in economic and political integration, as well as a means of enabling
agreement on the Single Market measures, it was to become clear
that it had far wider implications. As was argued in the introduction
to this chapter, the single economic space created pressures to create
a single political space. The limited majority voting which was
formalized in the Act also paved the way for acceptance of this
decision-making procedure in other areas of Community competence.

That the Single Market and the Single European Act were part of
the so-called 'Russian doll' strategy was soon revealed. In 1988 the
Delors Report was published which argued that further economic
and monetary union were necessary to complete the economic
reforms already in motion. It also set out the idea that there would
have to be three stages in the transition to full economic and
monetary union, an idea which was later to be incorporated into the
Maastricht agreement. The Delors Report was not launched into a
political vacuum; instead, there was a strong constituency in the
European Parliament, and amongst most of the member state
governments, which was receptive to the Report and enthusiastic to
extend what was already seen as the success of the 1992 programme.
By the time of the June 1989 summit, only the UK was opposed to
the Delors proposals and it was unable to prevent an intergovernmental conference being called to discuss further amendments to the
Community's Treaties so as to incorporate economic and monetary
union.

The UK government, which was already uneasy at the prospect of a deepening of the EC's economic space, was to be further alarmed by the next twist in the course of events. In April 1990 president Mitterand and Chancellor Kohl sent a letter to the President of the European Council proposing that a second intergovernmental conference on political union should be held in conjunction with that on economic union. The aims of the second conference would be to address some of the issues of the democratic deficit in the Community as well as the need for greater efficiency in its procedures. Once again, it was seen that Franco–German agreement and initiative were crucial to the historic leaps in European integration. This is not to argue that Mitterand and Kohl were pushing a reluctant Community along the road to greater union. On the contrary, the eventual success of their initiative depended on there being – as for economic and monetary union – a receptive constituency in favour of reform. The initiative was well received by many of the member states and by the European Parliament which had long advocated greater union. It was also welcomed by Delors, for whom it represented yet another 'Russian doll'. As a result, political union, which had been overshadowed by the Single Market programme during the second half of the 1980s, was now firmly back on the Community's agenda.

In the run-up to the intergovernmental conferences there was intense negotiation amongst the member states and the Commission. As a result of some critical compromises and some highly political and high-profile rephrasing of the draft text (for example, to remove the term 'federal'), it was possible for the heads of state to reach agreement at Maastricht in December 1991. The Maastricht agreement was based on three central pillars:

- *First pillar.* Existing treaties were to be extended in order to widen the scope of EC competences so as to include such areas as consumer protection, culture, education and development. The most important of these extensions was the agreement of a firm timetable for the previously agreed goal of economic and monetary union (EMU); this was justified as being essential to the Single Market. EMU was to be achieved in three stages

Table 6.1   The Maastricht Treaty programme for Economic and
Monetary Union

Stage I   1 July 1990
- currency linkage through EMS
- abolition of all controls over capital movements
- initiate convergence process towards four criteria:
  — inflation within 1.5% of three lowest
  — interest rates with 2% of three lowest
  — budget deficit not excessive
  — no downward devaluation over past two years
- completion of Single European Market

Stage II   1 January 1994
- creation of European Monetary Institute to monitor progress
- convergence monitored with greater vigour
- report on progress by December 1996 detailing EMU membership and
  timetable

Stage III   1 January 1999 (unless otherwise specified)
- creation of European System of Central Banks (ESCB) comprising the
  European Central Bank and the national central banks
- ECB and national banks to become fully independent
- ESCB objective is 'price stability'
- ECU becomes common currency in substitution for national currencies
- procedures for controlling budget deficits strengthened
- official foreign exchange reserves to be managed by ECB

(table 6.1) which would progressively lead to the closer
harmonization of the national economies, culminating in a single
currency and a European Central Bank:

(a) The first stage was deemed, retrospectively, to have begun in
July 1990. It involved currency linkage through the existing
Monetary System (EMS), abolition of controls on capital
movements, and the beginning of convergence of the
member states' economies. Convergence was a prerequisite
to passage to the second stage, and it was to be measured in
terms of inflation rates (within 1.5 per cent of the three
lowest rates in the Community), interest rates (to be within

2 per cent of the lowest three in the Community), the budget deficit (less than 3 per cent of GDP) and public debt (less than 60 per cent of GDP), and foreign exchange rate stability (no devaluations over the last two years).

(b) On 1 January 1994 the second stage could come into existence, involving more vigorous monitoring of convergence and the establishment of a European Monetary Institute to co-ordinate EMU.

(c) In January 1999 all those member states which had fulfilled the convergence criteria would enter the third stage, which was to be characterized by fixed exchange rates leading to a single currency and a European Central Bank. This timetable could be shortened. The heads of state were to decide by qualified majority, by no later than 31 December 1996, which countries were eligible for the third stage. If seven countries were ready, then the single currency and the central bank could be introduced as early as 1997.

- *Second pillar*.   It was agreed to establish a common foreign and security policy. This was to be organized on an inter-governmental basis outside of the Community's ordinary institutions. This was linked to recognition of the Western European Union as the defence arm of the European Community within NATO and, to this end, its headquarters were relocated from Paris to Brussels. However, there was considerable ambiguity as to precisely how the WEU would function and as to how the EC would develop its common defence policy. There was also considerable compromise over the mechanisms for determining foreign policy; common positions required unanimous support while implementation depended on qualified majority voting. The real issue, however, was the ill-defined definition of what constituted common positions as opposed to implementation. Partly in recognition of such difficulties there was also agreement that this second pillar would be reviewed in 1996.

- *Third pillar*.   There was agreement on greater co-operation over justice and home affairs with respect to immigration, asylum, drugs and organized crime. In particular, a common asylum policy was to be agreed by 1993, which was to be reviewed in 1996. This was very much an institutional adjustment to the enhancement of freedom of movement of persons, goods and capital within the EC space. It largely extended existing agreements in this area. Since 1975 EC member states had been co-operating over internal security via the Trevi system. Originally established to co-ordinate responses to the threat of terrorism, over time Trevi had become involved with issues such as asylum and illegal immigration. Maastricht was important in formally recognizing the competence of the EC in this area. It also formally recognized the 1985 Schengen agreement between Germany, France, Belgium, Luxembourg and the Netherlands which had abolished internal frontier border controls amongst the group. This third pillar was also to be handled on an intergovernmental basis outside of the Community's ordinary institutions.

In addition to the three pillars, a *Social Agreement* was also appended to the Maastricht agreement. This made it possible for the signatories to take qualified majority decisions in such areas as working conditions, and companies' disclosure of information and consultation with their workforces. This was an attempt to establish an EC social space, althouh it was enabling rather than prescriptive as very little of the actual contents of EC social regulation were specified in the Social Agreement.

The Maastricht agreement also addressed some of the issues of *decision-making* in the Community. It made three significant contributions in this respect. Firstly, qualified majority voting was extended to some of the new competences as well as to aspects of environmental policy. Secondly, a Committee of the Regions was established to increase the involvement of the sub-national state; its role was purely advisory but some commentators saw it as a further attack on the powers of the nation state. Thirdly, the European Parliament was granted a further extension of its limited powers. It was given a form of co-decision in some areas including environmental and research issues, internal market rules and free circulation of workers. Even in these areas of competence it was still largely restricted to having negative powers to reject measures which had been proposed by the Commission and the Council. The one advance for the Parliament was that if there was a difference of opinion between the Council and itself in one of these areas of competence, then a joint conciliation committee could be set up to try and reconcile the two bodies. Therefore, several small steps were taken to deepen the European political space but this remained dominated by the individual member states.

When the EC heads of state announced the Maastricht agreement it was acclaimed, by its supporters, as a major step forward in the process of European integration. Ross (1991, 65) argued that 'the Maastricht European Council of December 1991 was a clear sign that the construction of a new integrated and federalizing Europe had passed the point of no return'. The acrimonious debate which surrounded the agreement in many countries reinforced the idea that Maastricht was a fulcral point in the evolution of European integration.

A more detailed examination of the Maastricht agreement suggests that a less dramatic assessment of its importance may be appropriate. Firstly, two member states had not signed up for all of the Maastricht agreement. The UK opted out of the Social Agreement and had insisted that the whole of this element should be institutionalized as an opt-out by the other member states from the normal Community mechanisms. In addition, the UK had also opted out of the binding third stage of EMU, instead reserving the right to opt in if it so wished. Denmark also insisted on the right to hold a referendum on whether it would pass to the third stage of EMU. In effect, therefore, Maastricht formalized the idea of 'Europe *à la carte*' whereby member states could opt in or out of EC institutions and agreements. In this sense, it represented a defeat for the idea of an all-embracing Community. Secondly, it has been argued that there is little in the agreement which anyway was not possible under existing Community treaties (Usher 1992). Thirdly, Maastricht was stronger on broad objectives than on policy details in many areas including foreign and security policy and social policy. Decisions on many important issues had been postponed until the 1996 review. Even EMU was surrounded by ambiguities obscuring the precise dates of the transition phases and how closely a country had to adhere to the convergence criteria in order to proceed to the third stage.

When viewed in this light, the importance of Maastricht is seen to be its symbolism, for what it really represented was a shift in the balance between two very different views of the Community: the negative view that its role was to remove barriers (for people, trade and so on), and the positive view that Europe in the 1990s required supranational interventionism. As *The Financial Times* (3 December 1991) stated at the time: 'from now on almost anything can be considered of common interest'. This is precisely why it was derided by its critics as a charter for federalism and supranationalism. While there was very little of a direct federalist political nature in the agreement, they recognized that many of its measures, such as EMU and foreign policy, would generate pressures for federal political institutions to control its increasingly interventionist and supranational role. In this sense Maastricht was not the final 'Russian doll' within the Single Market programme but, instead, was just one more 'Russian doll' in the long chain of integrationist measures.

The Gathering Storm: EMU and Foreign Policy

December 1991 represented the culmination of a wave of optimism and dynamism in EC integration which had originated in the mid-1980s and led to the historic landmarks of the Single European Act and the Maastricht agreement. Over the following two years the road to European integration became far more difficult. This is explained by three features: the changing economic context for integration; difficulties over ratifying the Maastricht agreement; and the severe testing of the accord with respect to both foreign policy and EMU.

The first feature was that by the early 1990s much of Europe was in deep recession, and real GDP growth in the Community had fallen to only 1.1 per cent in 1992. Even more importantly, there was a realization that the deeper structural economic malaise of the EC, which the Single Market had been intended to remedy, was still considerable. Unemployment was higher and employment growth was lower in the EC than in its major rivals, the USA and Japan (figure 4.1). In addition, while intra-EC trade had increased sharply, the EC's share of trade with the rest of the world had fallen. The prognosis was equally unpromising; in March 1993, Mr Volker Jung, one of the directors of Siemens, argued that:

> The real question for me is where production will be ten years from now in world terms. The answer is East Asia, especially greater China – Hong Kong, Taiwan and mainland China. There you have the market plus low labour costs. Growth there will be much greater than in Europe, in terms not just of production but of product development.   (*Financial Times*, 9 March 1993)

By late 1992 concerns about both the immediate and the long-term outlook were such that promoting economic recovery in Europe became a major item on the agenda of the Edinburgh summit of the European Council. As a result the Edinburgh declaration suggested that member states switch public expenditure to growth-supporting programmes, encourage private investment, improve efficiency by reducing subsidies or enhancing competition, and attempt to achieve wage moderation in the public sector. Taken together, it was estimated that such national and EC measures could boost real GDP by 0.6 per cent by 1994 (Commission of the European Communities 1993a)

The recession had several important impacts on the process of integration. Even the publicly vaunted effects of the Edinburgh growth package were very limited, and this served to underline the limits to the EC as a supranational body intervening in a depressed capitalist economy. The recession also cast doubt on some of the hoped-for economic advantages from EC mega-initiatives such as the Single Market and EMU. Furthermore, it concentrated political leaders' minds on short-term domestic economic and political requirements. These severely eroded the positivism and confidence which had surrounded integration since the mid-1980s.

The second area of difficulty arose from the process of ratifying the Maastricht agreement. Although this was formally signed in February 1992 it could not come into effect until it had been ratified in all twelve member states. By June 1992 the agreement had encountered its first obstacle when it was rejected in a Danish referendum. Shortly afterwards it was ratified in referendums in Ireland and France, but the narrow majority in the latter further undermined confidence. Then in the autumn of 1992 the UK government was only narrowly able to secure parliamentary support for the initial approval of the agreement, and this was achieved by a series of political compromises including delaying its final parliamentary debate until the following summer. Much of the autumn period was then taken up by a damaging public argument between some member states and the Commission about the precise meaning of the Maastricht agreement and the future direction of integration. Eventually, the Edinburgh summit was able to offer some political support to the Danish government by allowing it exemptions from the third phase of EMU, from the common defence policy, from the agreement on common citizenship and from co-operation on justice and immigration matters. This further reinforced the idea that the EC was becoming an *à la carte* organization. However, it was sufficient to allow the Danish government to secure a yes vote in a second referendum in May 1993. This was followed by the approval of the UK legislature in July 1993, which completed the ratification process. However, while the agreement had now been approved by the member states, there had been a high cost to pay: some deep divisions had been revealed within the Community over the direction of future integration.

The third factor which undermined the momentum of the integration process was the difficulties experienced in implementing the Maastricht proposal. This was clearly illustrated during 1992 and 1993 by both common foreign policy and EMU.

There was a strong case for an EC foreign policy, based on the size and range of the Community's external economic relations, and its collective vulnerability to external shocks such as oil crises. The end of the Cold War also opened up the possibility of reduced American and increasing EC influence in Europe. Despite the commonality of broad interests, the context for the development of an EC foreign policy was unhelpful. Firstly, the member states had diverse traditions of foreign policy ranging from the neutrality of Ireland and Denmark to the second-rank superpower status of the UK and France (which had their own substantial military forces, permanent membership of the UN Security Council, and so on). Germany was in an ambiguous position; a unified Germany had the size and resources to be a major player in world affairs but its constitution did not allow German forces to be 'used outside the NATO area. In addition, there was the difficulty of disentangling a separate EC foreign and defence policy from that of NATO, still dominated by the USA.

It was not surprising, therefore, that the EC had steered clear of foreign policy involvement during much of its existence. A foreign policy mechanism, European Political Co-operation (EPC) had been established in the early 1970s, but it had very limited success except in terms of loose co-ordination of member states' foreign policies. Even though the role of the EPC was formally recognized in the Single European Act, it was essentially 'always intended to com-plement and coordinate national foreign policies rather than replace them' (Swann 1992, 48). As a result, the incorporation of common foreign and defence policies as one of the three pillars of the Maastricht agreements seemed to signal a major advance in the Community's role in this field.

Whatever optimism surrounded this notion was to evaporate quickly in the face of the crisis of the disintegration of Yugoslavia. This was brought to a head by the attempts of Croatia and Slovenia to break away from the Yugoslavia federation in 1991, followed by the tragic conflict in Bosnia. The EC had a strong and clear interest in a major crisis on its doorstep: one-half of Yugoslavia's trade was with the Community; the republics were potential future applicants for membership; there was a fear that ethnic conflicts could spread through the Balkans and involve Greece; and the resultant refugee crisis caused major political crises in Germany and Italy. However, the Community proved unable to respond to the crisis in a coherent and effective manner. This was evident in a number of ways. Firstly, there was Germany's early recognition of the independent Croatian and Slovenian republics, while other EC member states advocated a more cautious approach. Secondly, the attempt by the EC to act as the principal mediator in the crisis failed and it was replaced by overall UN co-ordination. Thirdly, there were discreet but continuous disagreements amongst the member states over the negotiations, sanctions and the form of external humanitarian and military intervention in the crisis. Although the EC had largely failed at the first major challenge to its post-Maastricht common foreign policy, this has to be seen in context of the severity of the crisis and the inadequacies of even the UN and USA responses, at a time when Community policy was still in the formative stage. An optimistic view would be that 'the progress towards a European identity in foreign and security policy will be limited and will develop only stage by stage' (Regelsberger 1993, 289). A more pessimistic view would be that the Yugoslavia crisis had exposed a significant 'capability–expectations gap' (Hill 1993, 306), and has raised the question of whether 'collective action can be *sustained* over time without a further leap into federalist obligations and structures' (p. 325). The evolution of foreign policy will depend on both external events and internal EC politics.

The Maastricht agreement also faced a major challenge over the three-stage timetable for and, implicitly, over the goal of EMU. Even in December 1991 there were some reservations about the EMU timetable which was considered to be too long, and about the convergence criteria which were seen as being too inflexible. However, this was perceived as being the price for persuading Germany to give up the Deutschmark. Therefore, there was a growing realization that very few countries might actually qualify to pass to the third stage of EMU; for example, in 1992 only Luxembourg strictly met all four targets for convergence, while France, Germany, the UK, Denmark and Ireland met only three of these. The Netherlands and Belgium met only two of the criteria,

Spain met only one, while Italy, Portugal and Greece failed all the tests. However, up to the middle of 1992 there was little sense of the gathering storm which would threaten to wreck EMU.

There were, strictly speaking, two interlinked storms which would break on EMU. The first of these was the weakening of the Exchange Rate Mechanism (ERM) whereby most EC currencies had been tied to narrow (2.25 per cent or 6 per cent) fluctuating bands around their central values. A combination of structural weaknesses in the economies, combined with interest rate inflexibility on the part of Germany and speculative attacks on individual currencies, led to progressive weakening of the ERM. The UK and Italy were forced out of the ERM in September 1992, followed by major devaluations of the Portuguese and Spanish currencies in November of that year, and of the Irish punt in January 1993. Then in the summer there was renewed pressure on the remaining currencies, especially the French franc, which led to the adoption of much wider 15 per cent fluctuation bands for the ERM. In effect, this signalled the end – if only temporarily – of exchange rate stability. There were a number of reasons for this crisis but probably the most important factor was the unexpectedly high costs of German unification. This had led to relatively high inflation rates and a public-sector deficit which, in turn, had resulted in the *Bundesbank* maintaining interest rates which were higher than could be sustained by other member states faced with economic recession. Figure 6.1 charts the movements of the principal currencies as the ERM disintegrated during this ten-month period. This alone meant that the convergence criterion of stable foreign exchange rates would become very difficult to achieve within the timetable for EMU.

The second storm was the depth of the economic recession which affected all EC member states by 1992 and 1993. This led to spiralling public-sector deficits and debt as governments were faced with the double squeeze of falling tax bases and rising social security spending. As a result, the position of all member states worsened with respect to government debt and/or borrowing (figure 6.2). There had, therefore, been divergence rather than convergence in terms of another of the key criteria for the third stage of EMU.

Exchange rates (DM per currency unit) rebased 12/1/87 = 100

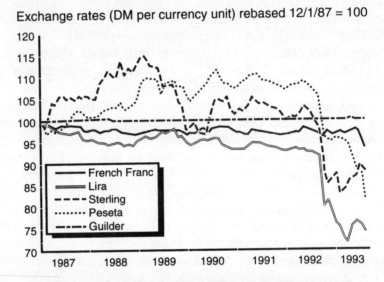

Figure 6.1   EMU and convergence: exchange rates, 1987–93
Source: European Commission

With hindsight, it would seem that the timetable set for EMU was unnecessarily rigid and demanding. EMU could probably be achieved without all countries having to meet all the criteria set down in the Maastricht agreement. Arguably, the key issue is not convergence but the setting of equilibrium exchange rates which will be tolerable for all the economies involved. Parallels can be drawn with German unification, where the exchange rate for the East German mark was set at such a high rate that it contributed to further increasing unemployment rates as companies were unable to compete against their more efficient West German counterparts. There is little doubt that equilibrium exchange rates will be critical when entering the third stage of EMU. However, the German experience also highlights the real cost of EMU, that is, the loss of economic sovereignty (over interest rates, exchange rates, monetary policy, and so on). At a time of recession this may imply the further depression of growth rates as individual countries lose the ability to

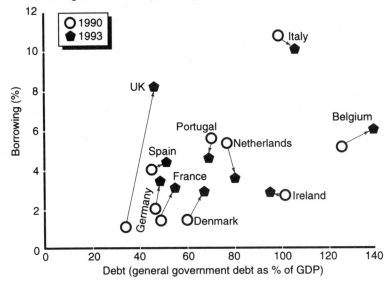

Figure 6.2 EMU and convergence: government borrowing, 1990–3
Source: European Commission

respond via strategies such as competitive devaluations or cutting interest rates to increase demand or reduce costs. There is nothing new in the arguments: all these possibilities were known at the time the Maastricht agreement was signed. However, the weakening of the ERM in 1992–3 did lead to a loss of confidence in EMU, and more generally in the cumulative nature of European integration. Subsequently there has been more questioning of whether these costs are politically and economically worthwhile.

SEVEN

The Contradictions of Integration

While the importance of politics in the shaping of the EC should never be underestimated, the central logic behind much of the long-term evolution of the Community has been the requirements of capital accumulation. The need to make EC capital competitive at the global scale has been the overarching aim and the organizing force behind the most significant shifts in EC policies, even if these have been given precise form by political necessities. This was evident in the emphasis given to the customs union, the common external tariff and the common markets in labour and capital in the 1960s. It was equally prominent in the emergence of a high-technology policy and the Single Market programme in the 1980s. More contentiously, it can be argued that economic necessity was the driving force behind the enlargements of the Community in the 1970s and 1980s. There were important advantages for EC capital in the extension of labour and capital markets and in the enlargement of product markets within the common external tariff, even if there were also significant political considerations.

It was inevitable that the largely capital-driven process of integration would engender a number of contradictions. There were sharp divergences of interest between member states, even with respect to economic logic. These were related to the balance of economic interests within each member state, and to its role within the European and the larger world economy. The interests of, say, Greece and FR Germany were bound to differ in terms of such issues as technology policy, capital liberalization and removal of non-tariff barriers. Northern and Southern European member states also differed in their relationships with non-EC Mediterranean states.

The relationship of each national economy to international capital also differentiated states. Hence, the UK and Ireland, which received substantial inward investment from outside the EC, were more receptive than, say, Italy and France to American and Japanese transplant operations.

The contradictions of integration run much deeper than a conflict between the divergent interests of member states. Even in terms of economic goals, the Community has been faced with incompatible objectives: there has been an uneasy relationship between fostering competition within the EC and the need to develop national or European champions able to compete on the global scale. Regional economic goals have also clashed with the overall growth aims of the Community. In addition, there are other, more wide-ranging contradictions. Political, social and environmental goals often conflict with the economic goals of the EC; and there are a number of sharp inconsistencies between the specific objectives being pursued within each of these larger goals. This chapter reviews some of these major contradictions.

Transnationals, Competition, European Champions and the State

There has been strong growth of transnational companies (TNCs) in the capitalist economies during the post-war period. This is based on the needs to obtain scale economies (especially in industries with high R&D costs), reduce production costs and secure market access. Given the national segmentation of the Western European economy – compared to, say, the USA – it is not surprising that the TNCs are especially active in the EC. Of the estimated 4500 TNCs in the global economy in the late 1980s, 2500 were based in the EC and generated $750–850 billion in revenue (Tempini 1989). In a way, this represented fulfilment of one of the original aims of the Community: the creation of large companies capable of competing internationally. Indeed, this has been one of the principal aims of the 1992 programme: to create scale economies so as to increase efficiency and, in the longer term, generate greater potential for R&D, especially in high-technology industries.

While the strengthening of international capital fulfils some of the aims of the EC, it also conflicts with another objective, that of maintaining competition within the Community. Indeed, this is one of the cornerstones of the Treaty of Rome. The problem is that, of the 2500 TNCs in the EC, just 450 dominate fifteen crucial industrial sectors (Tempini 1989). There is another complication in that many of the largest TNCs in the EC – such as IBM in computers, Ford in cars, Hoffmann-La Roche in pharmaceuticals – do not have their headquarters in the Community. Not only is there the possibility that TNCs may diminish competition in the EC, but this may be accompanied by the loss of production and market shares to externally owned capital.

The 1960s was probably the critical decade in inward investment of (largely American) capital to the EC. However, this is a continuing process. Between 1982 and 1986, Europe's 1000 largest corporations were involved in 256 international mergers and acquisitions; 36 per cent of these involved links with companies outside the Community (*Financial Times*, 5 October 1987). At times, EC companies were the dominant partners in these deals but, nevertheless, there was continuing inward investment into the Community (see also table 5.7a).

The growth of transnational capital also represents a challenge to the power of the state. TNCs are locationally flexible and have considerable bargaining power *vis-à-vis* national governments. There are many instances of TNCs overriding the priorities of national governments or winning major financial concessions from them. Examples include Ford's decision to locate new plant in Portugal and IBM's challenge to the UK's regional policies. However, in the 1980s there has been growing awareness that TNCs also represent a serious challenge to the powers of the EC and, indeed, could undermine some of the aims of the 1992 programme. Not surprisingly, EC attitudes to TNCs have been 'fundamentally two-edged, on the one hand encouraging multinational activity in transnational European markets, while on the other trying to remedy the disadvantages and concerns caused by MNEs' [multi-national enterprises'] activities' (Tempini 1989, 15).

These contradictions have been highlighted by the contrast between the Single Market emphasis of the 1992 programme and the potentially anti-competitive behaviour of TNCs. Another problem has been the existence of divergent national legislation governing TNCs' activities. There are important differences between the member states in the rules governing competition, taxation, information disclosure and takeovers. Combined with the lack of a coherent EC framework for the regulation of international mergers and acquisitions, this could lead to national segmentation of the supposedly homogeneous economic space of the Single Market.

The EC has not been completely powerless in regulating TNC activities. Articles 85 and 86 of the Treaty of Rome prohibit 'all agreements between undertakings, decisions by associations of under-takings and concerted practices' which restrict the free interplay of competition (Commission of the European Communities 1985b). In practice, EC powers were largely limited to price or market fixing, or other abuses of dominant market positions, where these affected intra-member trade. Where companies infringed Articles 85 or 86, the Commission had the power to order a cessation of such practices. The Community could respond to formal or informal complaints about such anti-competitive behaviour, or initiate its own inquiries into particular sectors; the brewing and the margarine industries were the first to be subject to this process. If persuasion failed, then, as a last resort, the Commission could impose fines and/or refer cases to the European Court of Justice (Cini 1993).

Articles 87 and 235 of the Treaty of Rome also gave the Commission power to review international mergers between TNCs if they involved companies with turnover in excess of 1000 million ECU. In practice, there was little attempt to pursue EC policy in this area until the late 1970s when Commissioner Davignon assumed responsibility for industry. However, the powers of the Commission were limited, not least in that they could only intervene after a merger or acquisition had occurred. With the onset of the 1992 programme, there was an urgent need to review these powers.

In 1989 it was proposed that member states should yield to the EC
power to intervene in international takeovers; until 1992 this would
apply to takeovers involving companies with a combined world
annual turnover in excess of 5 billion ECU, and thereafter the
threshold would be lowered to 2 billion ECU. Negotiations were
difficult: the larger EC states, such as France, FR Germany and the
UK, thought the financial thresholds were too low, while the smaller
states considered they were too high. France was also unhappy with
the sole reliance on anti-competitiveness criteria, and argued that
exceptions were necessary so as to allow the EC to develop
'European champions' capable of competing against the USA and
Japan. This, of course, was a manifestation of the old EC dilemma of
competitiveness versus competition. The eventual compromise,
agreed in December 1989, was to give the Commission power to vet
mergers and takeovers which involved companies with a world
turnover of 5 billion ECU, if a minimum of 250 million ECU of each
company's sales were located within the EC.

These powers were reviewed in 1993 when the Commission
proposed lowering the turnover threshold above which takeovers are
automatically vetted by Brussels. In futures, this would have
encompassed future significant mergers, such as that between the
Reed and Elsevier publishing companies in 1992, which had
previously slipped under the threshold. However, this was opposed
by Germany, France and the UK on the grounds, amongst others, of
subsidiarity.

These measures were likely to increase the powers of the state, or
the suprastate, versus international capital. However, they did little
to harmonize company regulations across the Community. The
Commission did produce a draft directive to harmonize takeover
practices in the Community, but it was opposed by a number of
states. In this respect the EC economic space remained highly
fragmented. National regulations concerning takeovers and mergers
are an especially sensitive area of policy, which has led to criticisms
of a lack of 'a level playing field'. The UK has a relatively liberal set
of regulations compared to most other major EC states. Not
surprisingly, 73 per cent of acquisitions in the EC by value in 1988

were in the UK (*Financial Times* 27 November 1989). Even more striking was the fact that twenty-three out of the twenty-six successful *hostile* international takeover bids in the EC were located in the UK. There are three main reasons for this. Firstly, there are legislative differences but, in the medium term, the development of a European company statute should eliminate these. Secondly, except for the UK, the Netherlands and FR Germany, most companies in the other member states have placed less than half of their shares on the open market. Hence, they are relatively secure against hostile takeovers. In the longer term, it is possible that this part of the 'playing field' may be levelled by EC company regulations. However, there is a third and deeper difference rooted in national business cultures. Only in the UK is management fully accountable to shareholders and hence under pressure to accept the highest share-price offers, even if hostile. This is partly related to the disproportionate influence of the City of London in the UK economy. Elsewhere in the Community, longer-term business targets or the interests of employees may be accorded higher priority, hence increasing resistance to takeovers. Given these differences it appears rather strange that the UK was the strongest opponent of the EC draft directive on harmonization. However, this is explained by UK preference for a voluntary takeover code rather than a legally binding code. It reflects the belief that, at a cultural level, a European company statute is unlikely to guarantee a level playing field between countries. It may also mean that some of the economies of scale and the competitiveness objectives of the 1993 programme – which are based on mergers and takeovers – are, in reality, unobtainable.

By the early 1990s there had been a further shift in the debate between the proponents of concentration and of competition. Against a background of deepening recession and renewed crises in industries such as textiles, steel and shipbuilding, there was growing dissent against what had become the neo-liberal orthodoxies of the 1980s. Sadler (1992, 1726) writes: 'This suggested that it was necessary to recast the debate around industrial policy, and to ask again some old questions in a new environment.' The old question

Table 7.1   State aid to manufacturing in the EC, 1986–90

	ECU per person employed (annual averages)		ECU million (annual averages)	
	1986–8	1988–90	1986–8	1988–90
Belgium	1,606	1,655	1,175	1,211
Denmark	593	634	316	333
Germany	994	984	7,869	7,865
Greece	2,983	1,502	2,074	1,072
Spain	1,749	936	4,491	2,499
France	1,437	1,380	6,479	6,106
Ireland	2,114	1,734	447	368
Italy	2,139	2,175	10,760	11,027
Luxembourg	988	1,270	37	48
Netherlands	1,215	1,327	1,101	1,225
Portugal	302	758	245	616
United Kingdom	770	582	4,101	3,133
EC Total	1,325	1,203	38,835	35,503

Source: European Commission

was whether Europe needed a small number of major companies that could hold their own in the global market. In theoretical terms this can be analysed according to the alleged dynamic effects, such as product innovation, which are fostered by a more concentrated industrial structure; in other words, that effective R&D requires large companies able to operate at the global scale (Jacquemin and Slade 1989).

Amongst the member states the argument was taken up by France and Italy in terms of the need to support 'European champions'. This is not surprising, given the tradition of state intervention to create 'national champions' in both of these economies. As can be seen from table 7.1, both countries have relatively high levels of state aid to the manufacturing sector in comparison to countries such as the UK, Denmark and Germany. It was also significant that the first case in which the EC vetoed a proposed merger, under its new powers, involved companies from these two countries. Aerospatiale and Alenia were refused permission to take over the Canadian company, de Havilland. The reasons for this were illuminating. It was not so much the fact that the new company would control 63 per cent of the EC market which was important, as the fact that it would also have 50 per cent of the world market for small passenger airlines. This underlined the importance of examining competition issues for other EC companies at the appropriate scale for that particular sector.

The arguments over the de Havilland takeover brought to a head the political tensions amongst member states which were accustomed to state-led monopolies, such as France and Italy, and those which favoured a *laissez faire* approach, such as the UK, Germany and the Netherlands. Similar tensions existed within the Commission between more interventionist commissioners, such as Bangermann, and free-marketeers such as Brittan. The shift was evident in the statement of Karel van Miert, the competition commissioner: 'I very much believe in competition, but not as a religion. You must observe a proper balance' (quoted in *Financial Times* 26 March 1993, 19). In fact, the argument between competition and competitivity can be restated as one about the scale at which competition rules apply. The argument concerning European champions can be seen as a counter-strategy, given that there is a lack of a global framework to regulate competition and ensure that there is a world-wide level playing field. In other words, 'deep integration' (Jacquemin 1993) needs to be applied at the world scale, for there are significant economic constraints on the autonomy of the EC in regulating competition.

Budgetary Crises: the Long Shadow of Agriculture

The media often present the operations of the EC as a series of recurrent budgetary crises. There is an element of truth in such a portrayal, for the range and the form of EC policies are influenced by the constraints on its financial resources. This section examines the nature of these crises, which are rooted in a persistent failure to match income and expenditure.

There have been important changes in the way in which the EC is financed. Until 1970 there was a system of annual national subventions, but thereafter the Community was given its 'own resources'. These were derived from three sources: customs duties levied at the frontiers of the EC; agricultural levies on food imports; and a proportion of national VAT receipts which was transferred to the Community, subject to a maximum of 1 per cent. This was a simple but a crude and inherently inegalitarian system: it penalized smaller states such as the Netherlands which had large international ports serving the EC as a whole; and it favoured richer states where VAT tended to be a smaller proportion of GDP, as consumption was a smaller proportion of national income (Shackleton 1989). In fact the two states which carried the largest burdens were the UK (19.5 per cent of contributions, but receiving only 12.6 per cent of expenditure by the EC in 1985) and FR Germany (28.2 per cent and 17.0 per cent respectively). Germany accepted such unequal transfers, given its national prosperity and the advantages that the common market offered its industries. The UK was unwilling to do so and, as was noted in Chapter Four, this led to a series of political conflicts in the 1980s. These were resolved at the 1984 Fontainebleau summit which guaranteed the UK a rebate of two-thirds of the difference between its contributions to and receipts from the EC.

An important reason for the concessions made to the UK at Fontainebleau was the need to reach unanimous agreement on raising the ceiling for the EC's 'own resources' from VAT receipts. This had become necessary because of increasing expenditure demands and because of the expected net transfer of resources to Spain and Portugal, following their accession in 1986. Although a new ceiling was agreed of 1.6 per cent as of 1988, expenditure continued to rise faster than expected and the new threshold was soon exceeded. The negotiation of a further new financial ceiling involved long and difficult bargaining. Eventually, a new ceiling was set at 1.2 per cent of the EC's GDP. This was to be achieved from the three existing 'own resources', but any shortfall was to be met by 'a fourth resource': direct national levies set in relation to member states' GDPs. This had provided some relief for the Community in

Table 7.2   Total budgetary expenditure in the EC, 1958–91

	Million ECU	5-year % increase	% spent on agriculture (EAGGF)
1958	35	—	0
1963	267	662.8	0
1968	2,624	882.8	85.7
1973	4,937	88.1	76.3
1978	12,190	146.9	78.3
1983	25,817	211.8	63.2
1988	41,804	61.9	68.7
1991	59,369	—	57.3

Source: Commission of the European Communities, *European Economy, Annual Economic Review*, no. 29, July 1986, 166; Shackleton (1989); European Communities (1992) *Financial Report 1991*

terms of income-raising, but the overall financial position is still under pressure from a rapid growth in expenditure. Decoupling the own-resource ceiling from VAT and linking it to GDP was critical in two ways; firstly, because of the propensity to save in richer countries, VAT receipts had not kept pace with GDP growth; and secondly, VAT receipts were regressive, unduly burdening both lower-income countries and individuals (Pinder 1991, chapter eight).

The EC's total budgetary expenditure increased from 35 million ECU in 1958 to over 59 billion ECU in 1991 (table 7.2). There were exceptionally high five-yearly rates of increase in the first decade of the EC's existence. In part, this simply reflected the necessary establishment of institutions and the early development of policies. It also reflected the creation of the costly CAP which, in 1968, accounted for 85.7 per cent of all expenditure. The rate of overall expenditure growth fell in the late 1960s, but increased substantially in the 1970s and early 1980s. This was in response to the first and second enlargements and to increased expenditure on regional, social and industrial policies. The European Regional Development Fund (ERDF) was created largely in order to counterbalance the

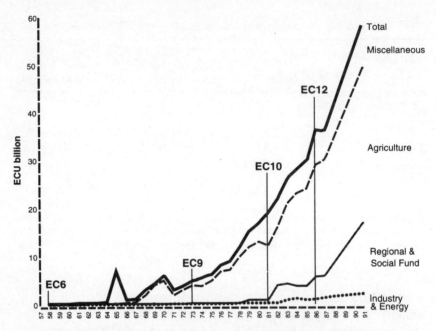

Figure 7.1   Development of the EC budget, 1958–91
Source: Commission of the European Communities

UK's expected net contribution to the EC budget, while the Integrated Mediterranean Programme was introduced to counterbalance the existing bias in many EC policies in favour of northern member states. In the mid-1980s expenditure growth slowed as the Community persistently encountered the legal ceiling on its expenditure. This was reinforced by the changing political climate for public expenditure in the 1980s. Most EC governments, led by the UK, attempted to reduce government interventionism and to shift resources from the public to the private sector.

Over time, new policy areas have made increasing demands on Community expenditure, but agriculture has remained dominant (see figure 7.1). In absolute terms, expenditure on the CAP increased from 4.5 billion ECU in 1976 to 21 billion ECU in 1986 (Gardner 1987). While the proportion of the budget devoted to agriculture has fallen, it still accounted for 57 per cent in 1991. In order to understand the EC's budgetary crises it is necessary to examine the CAP in more detail.

The CAP was established to bring market stability, higher incomes, modernization and low consumer prices to farming in the EC. It has brought a number of benefits (Lintner 1989a):

- Agricultural self-sufficiency has increased from 91 per cent in the EC6 in 1958 to 108 per cent in the EC10 in the early 1980s. This has contributed positively to the EC's global balance of trade.
- Real agricultural incomes have increased over time. While they have lagged behind the total increase in EC incomes, they have been more stable than farm incomes in, for example, the USA.
- There has been modernization of agriculture, with a substantial increase in capital–labour ratios. However, this is partly due to more general modernization processes in capitalist agriculture, as there have been similar tendencies in other non-EC economies (Williams 1987, chapter 4).
- It has been argued that the CAP, as the most important EC common policy, has contributed to the process of integration. While this was probably true in the 1960s and 1970s, it was more debatable in the 1980s when budget crises contributed to a failure to develop new policies and to stagnation in collective decision-making.
- Growing self-sufficiency has led to a threefold increase in EC agricultural exports, 1973–86 (Harrop 1989). Between 1979 and 1984 the EC's share of world exports increased from 12 per cent to 17 per cent. However, EC exports are subsidized by the Community and this has contributed to persistent trade disputes with other major exporters, especially the USA.

While the CAP has brought significant advantages to the Community, it also has a number of important disadvantages (Lintner 1989a):

- Food prices have tended to be higher in the EC than in world markets. This can be traced back to the political compromise between FR Germany and France which was necessary to launch the CAP in the 1960s (see pp. 47–9).
- The system of guaranteed prices and intervention purchasing has required heavy expenditure to store and dispose of the notorious lakes and mountains of surplus wines and foods. In 1987 the EC had an estimated 12.3 billion ECU of food in storage but deterioration meant that its real value was only one-third of this amount (Harrop 1989). By the late 1980s, CAP expenditure was equivalent to 49 per cent of the final value of agricultural output in the EC, which was substantially higher than the 22 per cent in the USA (*OECD Observer* 1988, 10). This meant that the CAP realized a transfer of about 1 per cent of the EC's GDP to the agricultural sector (Demekas et al. 1988).
- The net result of the CAP is to effect substantial income transfers from consumers to farmers and from the rest of the world to the EC.
- The Third World has suffered from the trade diversion effects of the CAP.
- Agricultural intensification and commercialization has often been accompanied by negative environmental effects.
- The existence of MCAs has seriously distorted what is supposedly a single market for agricultural products within the EC.
- While the CAP has increased farm incomes, the impact has been profoundly inegalitarian. Large, efficient capitalist farmers have benefited most while smaller-scale farmers have benefited least. This also has a regional manifestation, with the largest economic gains falling to northern EC farmers (see figure 7.2).

Given this catalogue of economic and political disadvantages, it is not surprising that the CAP has been heavily criticized. As Fennell writes:

The imbalance of supply and demand, the accusations of protectionism at home and dumping abroad, the high unit costs of subsidies, the

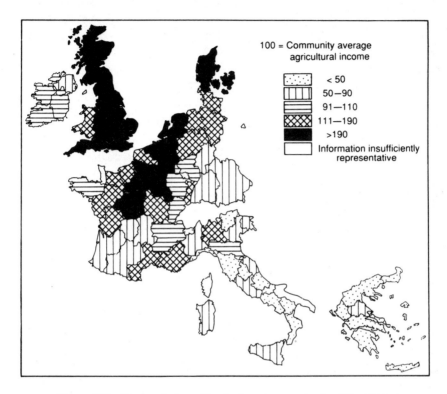

Figure 7.2   Regional disparities in farm income in the EC, 1981–2
Source: Lintner (1989a, 12)

persistently low incomes of many farmers combined with over-compensation of others, which leads to perverse income redistribution – all are economically embarrassing. (1987, 67)

There were attempts to reform the CAP almost as soon as the policies had been formulated. The best-known was the Mansholt Plan (Commission of the European Communities 1968) which proposed measures to concentrate output in larger and more specialized farms, while withdrawing some land from production. This was too radical for the expansionary 1960s and early 1970s, and it failed. Nevertheless, there was constant pressure on the CAP budget and, despite the well-organized opposition of famers' groups, some reforms were achieved. For example, real EC milk prices were cut by 1.5 per cent per annum, 1975–81 (*OECD Observer* 1988). However, this was inadequate in the face of two major structural problems. Firstly, output increases of 2 per cent a year were considerably outstripping an inelastic demand which was only rising at 0.5 per cent. Secondly, even price cuts of 1.5 per cent per annum were insufficient to cut supply when annual productivity gains of 2 per cent ensured that farmers' real returns continued to improve. Domestic political requirements, however, especially in France and FR Germany, ruled out the more radical reforms needed to overcome these structural difficulties.

The political will to break this policy impasse emerged only after a severe Community budgetary crisis in 1979–80, combined with a fall in world commodity prices in the following year. This led to a search for more radical reforms, and *Reflections on the Common Agricultural Policy* (Commission of the European Communities 1980) recommended the introduction of producer co-responsibility. In 1981 guidelines were agreed for limiting future food production, but they did not take into account existing surpluses. There was greater progress in 1984 with the agreement, in principle, that there should be no new MCAs while existing ones would be dismantled. More importantly, quotas were introduced in the same year as a means of limiting milk production. Yet expenditure on the dairy sector actually increased by 33 per cent, 1983–6. This was because EC Agricultural Ministers have 'consistently cheated in the application of the milk quota legislation so as completely to subvert its spirit and intention' (Gardner 1987, 168).

In 1986 stiffer dairy and beef quotas were introduced but these were still inadequate for the task; for example, the milk quotas still allowed 12 per cent more production than was required for EC sufficiency levels. The failure of such reforms can only be understood in context of the politics of the CAP. Firstly, ten of the twelve member states are net exporters of farm produce. They therefore benefit substantially from the Guarantee budget and from the system of export premiums and import levies. Secondly, although Germany is a net food importer, German farmers are well-organized and exert considerable electoral pressure in a number of important marginal constituencies. Hence, Germany has continued to be one of the major opponents of price cuts and, for example, vetoed a 1985 proposal to reduce cereal prices. The UK is also a net food importer but has been less reforming in practice than government rhetoric suggests. There are a number of reasons for this: there is growing self-sufficiency in the UK, and the National Farmers Union is a very effective pressure group. Furthermore, many Ministers of Agriculture have concluded that the CAP is unreformable and have therefore concentrated on expanding British output so as to maximize the benefits received from the farm policy (Gardner 1987, 178). Finally, all individual governments have been able to protest their commitments to reform while hiding their inaction or acquiescence behind the process of collective decision-making.

Despite these political obstacles, the perilious state of EC finances has forced some reform of the CAP. The 1988 Brussels summit was particularly important in this. It proposed that the growth of expenditure on guaranteed prices should not exceed 74 per cent of the growth in the Community's GDP. It also introduced budgetary stabilizers to limit expenditure where necessary. Together, these were supposed to exert effective budgetary discipline on the agricultural ministers for the first time (Pinder 1991, chapter five). Agricultural reform has also been boosted by the growing influence of the environmentalist movements. There is increasing pressure to reduce the use of pesticides and fertilizers, to pay more attention to protecting fauna and flora, and to 'set-aside'. The latter involves farmers taking land out of production in return for a subsidy. This is linked to measures to diversify farm incomes so as to reduce their dependence on food products for which the Community has large surpluses. The attraction of such ideas is that they link budgetary reforms to politically popular environmental policies (see Marsh 1989).

The Uruguay round of GATT negotiations has generated

additional pressures for reform of the CAP. As part of the overall agreement to reduce tariffs on manufactured goods and services, the EC has been pressurized into accepting agricultural reforms. These reforms, known as the MacSharry proposals, have two distinctive elements. Firstly, there are significant reductions in market support prices for cereals, which will bring prices far closer to world levels. Secondly, all farmers are to be compensated for lost production by direct payments. While there is a fixed amount of compensation per ton, yields are to be calculated on the basis of regional averages. In addition, the compensation payable to larger farms is subject to a 15 per cent set-aside requirement. Compensation will be paid for lost production but only up to fixed limits; for every 85 hectares on which compensation is paid, 15 hectares have to be set-aside.

The reform measures are expected to lead to substantial cuts in production, although it has to be emphasized that they excluded the troublesome wine, dairy and sugar sectors and offered only limited reform of the beef sector. Their effectiveness will also depend on how farmers respond to the changes in incentives. In addition, the reforms will have important redistributive effects (Josling and Mariani 1993). The price cuts and related compensation will be regressive, reinforcing the concentration of CAP expenditure in the more prosperous farming regions. However, within all regions farmers with less-productive land will be favoured, given that compensation will be calculated on the basis of average regional yields. The set-aside programme will tend to penalize larger farms and more efficient agricultural regions because of the maximum limit on the amounts of compensation available. In May 1992 the Council of Ministers accepted the MacSharry proposals after a long and protracted public dispute between member states, commissioners and farming organizations.

The MacSharry reforms were precursors to a US–EC accord on the agricultural element of the Uruguay GATT reforms. The key elements were the following (Swinbank 1993):

• phasing-in of new arrangements over six years;

- reduction of total support to agriculture by 20 per cent, compared to 1986–8;
- direct acreage and headage payments excluded from the calculations, to the benefit of European farmers;
- reduction of border protection measures by an average of 36 per cent. However, the EC is allowed to add a variable element to the tariffs if world prices fall by more than 15 per cent below 1986–8 levels;
- import opportunities to be opened up to allow 5 per cent of the domestic market to be captured by the end of the period;
- spending on export subsidies to be reduced by 36 per cent over the period, while the volume of such exports is to be reduced by 21 per cent.

Opinions of the reform package were mixed, the consensus being that it was acceptable, given that a failure to negotiate an agreement could have undermined the whole of the GATT negotiations. However, the arguments over the package continued until well into 1993, France being particularly unhappy over price levels. It is unlikely that the reforms will be assured until all the GATT reforms have been completed and implemented.

In earlier decades, any suggestion of weakening the Community's only major common policy would have been rejected out of hand. It is a sign of change in the EC, partly engendered by the 1992 programme and by the need for resources to develop a positive industrial policy, that such a reform is now considered possible. As Avery writes, in the 1970s there were:

> fears that changes in a 'cornerstone' of the EC could put at risk the whole construction of the Common Market and other common policies. Indeed the very term 'reform of agricultural policy' was taboo in official European language in the 1970s. Today it is non-reform of the CAP that is seen as an obstacle to progress in the European construction.   (1987, 162)

It is an obstacle because it lies at the very heart of the EC's budgetary crises and its ability to develop as a supranational organization.

Regional Inequalities:
the Inherent Contradictions of Growth?

The Spaak Report – which predated the Treaty of Rome – did not anticipate that any long-run disruptions or imbalances would result from the creation of a customs union. Instead, there was a belief that the 'invisible hand' of market forces would correct any significant regional disparities in a homogeneous economic space. Hence, when the Treaty of Rome was drafted, the 'emphasis on the virtues of the free market led it to give the Community only a secondary role in overcoming divergence, which it regarded as being largely temporary and exceptional rather than endemic and which could be minimized by careful co-ordination of national economic policies' (Hodges 1981, 44).

These essentially neo-classical and Keynesian economic analyses proved unduly optimistic. Firstly, the development of a common market, however imperfect, reinforced the advantages of central locations within the EC. Peripheral regions were attractive to some industrial sectors, such as textiles or car assembly firms in search of low cost labour. However, the more centrally located regions held important advantages in terms of accessibility, market potential, access to capital markets and R&D. Secondly, there was imperfect mobility of capital and labour within the Community so that structural rigidities were built into the pattern of regional inequalities. Thirdly, even if there were processes of convergence within the Community, they were counterbalanced by the impacts of the three enlargements. Each of these added a new member state – successively, Ireland, Greece and Portugal – which was bottom of the EC league table of per capita incomes at the time of accession. For example, average per capita incomes in Portugal were only one half of those in the Community in 1986, and none of the Spanish and Portuguese regions had incomes greater than the EC average (Kowalski 1989).

The measurement of regional inequalities in the Community is problematic. At a general level there is clear evidence that regional inequalities are considerably greater than in the USA; for per capita incomes the ratio is twofold while for unemployment it is threefold (Kowalski 1989). There is also evidence that the peripheral regions of the EC have consistently had lower incomes per capita, higher unemployment, greater dependency on (slow growth) agriculture and disproportionate representations of low-technology and low-growth industries. More specifically, in 1985 the ten most prosperous regions of the EC had per capita incomes which were three times greater than those in the ten poorest regions (Keeble 1989). The distributions of both GDP per capita and unemployment display marked regional variations (figure 7.3). The highest levels of GDP per capita are to be found in the UK, Denmark, France, northern Italy, the Netherlands, Belgium, Luxembourg and Germany. In contrast, the lowest levels of GDP per capita are found in Greece, Spain, Portugal, Ireland and southern Italy. The measurement of unemployment is more problematic but Community statistics suggest that the highest levels are in the northern UK, southern Italy, southern France, and especially Ireland and Spain. The latest term used to describe the economic centre of the EC is 'the hot banana', which refers to the arc running from the traditional axis of London – Paris – Amsterdam through southern Germany to Milan, before spreading through southern France to Barcelona/Valencia in Spain.

While it is possible to identify a static picture of deep-rooted regional inequalities in the EC, the data on regional convergence and divergence trends are more confused. Over the long period 1950–79, Molle (1980) has identified an overall trend of decreasing regional disparities. Jensen-Butler (1987) also found that, in the 1970s, there was a decrease in both international disparities and in inter-regional differences within most EC countries. However, by the late 1970s conditions seemed to be changing and Clausse et al. (1986) estimated that regional disparities in industrial output per capita were static between 1970 and 1982.

With the industrial recession in the early 1980s, there was another shift in the regional pattern. The older-established industrial regions were most severely affected by the process of industrial

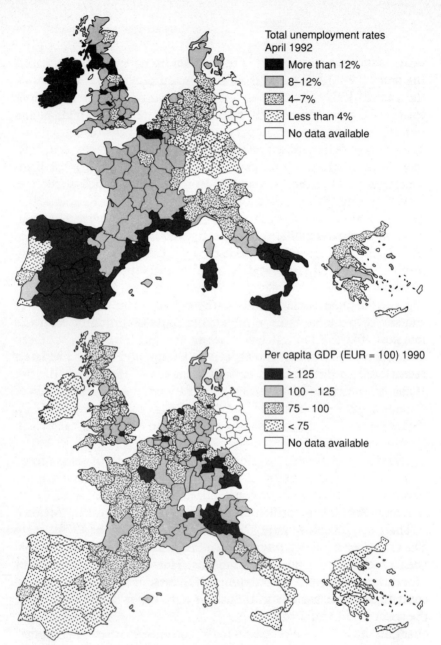

Total unemployment rates April 1992

- More than 12%
- 8–12%
- 4–7%
- Less than 4%
- No data available

Per capita GDP (EUR = 100) 1990

- ≥ 125
- 100 – 125
- 75 – 100
- < 75
- No data available

Figure 7.3   Unemployment rates (April 1992) and GDP per capita (1990) in the EC regions
Source: Eurostat

restructuring. Keeble's (1989) analysis clearly demonstrates that, in the period 1979–83, the older-established 'inner central' regions of the EC suffered the largest manufacturing losses. They were also least effective in attracting new service sector jobs (table 7.3). At the same time, the outer peripheral regions were the only ones to record an increase in manufacturing jobs. This aggregate regional shift was the outcome of several complex processes including productive decentralization (see Mingione 1983; Williams 1987, chapter 5), flows of inward MNC investment to peripheral regions (Hamilton 1987), and increased locational flexibility based on developments in information technology (Gillespie 1987; Jensen-Butler 1987). Molle (1990) confirms that there was a significant decrease in the overall levels of regional disparities of wealth in the EC in the period 1980–5.

A less optimistic picture is presented by an analysis of the early and mid-1980s (Commission of the European Communities 1987a); this shows that, after 1980, regional disparities were essentially static in both the EC10 and the EC12. Furthermore, Keeble's evidence reveals that, whatever marginal adjustments occurred in the early 1980s, the absolute differences in levels of GDP per capita were still considerable. In 1983 the mean GDP per capita in the inner central regions was almost double that in the outer peripheral regions of the Community. Perhaps even more pessimistic is Cheshire's (1992) conclusion, based on an analysis of functional urban regions, that there was marked regional divergence during 1971–88: the only important exceptions to this were regions such as Rotterdam and Barcelona which had had effective territorial policies.

There are several difficulties in interpreting regional trends in the EC. One of these – the effects of the three enlargements – has already been referred to. Another important feature is the changing nature of regional disparities. Spatial inequalities in the Community contribute to and reflect social and economic inequalities. Over time the spatial manifestations of underlying structural inequalities have changed. This complexity was recognized by the Commission in the 1970s; it identified five main types of disadvantaged region in the EC (Armstrong 1989):

Table 7.3 European Regional Development Fund national allocations, 1975–93

| | Quota allocations (%) | | 1985 range | 1986 range | Indicative allocations 1989–93 | |
	1975	1982			Objective 1 Regions	Objective 2 Regions
Belgium	1.5	0	0.9–1.20	0.61–0.82	–	4.3
Denmark	1.3	1.3*	0.51–0.67	0.34–0.46	–	0.4
FR Germany	6.4	0	3.76–4.81	2.55–3.44	–	8.9
France	15.0	2.4**	11.05–14.74	7.48–9.96	2.1	18.3
Greece	N/A	16.0	12.35–15.74	8.36–10.64	16.2	–
Ireland	6.0	7.3	5.64–6.83	3.82–4.61	5.4	–
Italy	40.0	43.7	31.94–42.59	21.62–28.79	24.5	6.3
Luxembourg	0.1	0	0.06–0.08	0.04–0.06	–	0.2
Netherlands	1.7	0	1.01–1.34	0.68–0.91	–	2.6
Portugal	N/A	N/A	N/A	10.66–14.20	17.5	–
Spain	N/A	N/A	N/A	17.97–23.93	32.6	20.7
UK	28.0	29.3	21.42–28.56	14.50–19.31	1.7	38.3

* Greenland only
** Overseas departments only

- underdeveloped regions, as in much of the Mediterranean area, tending to be dependent on agriculture, and having low incomes and poor infrastructure;
- declining industrial regions, as in the north of the UK and Wallonia;
- peripheral regions, such as Ireland, with poor accessibility to product markets;
- border regions – most notably between West and East Germany, until 1990 – where there are major barriers to trade with adjoining areas;
- urban problem areas, such as Naples or Belfast, which have severe social, environmental and economic difficulties.

In the early years, with the emphasis on growth and integration in the Community, regional problems were conceived of in terms of underdevelopment and were typified by the Mezzogiorno. Later, especially after the accession of the UK and following the industrial recessions of the 1970s and the 1980s, attention shifted to the problems of older industrial regions. By the 1980s, urban areas featured prominently in the analysis of spatial problems in the EC. In fact, there are at least two main types of urban problem area in the EC (Cheshire et al. 1986). Rapidly growing urban areas, such as Naples and Athens, have severe problems of congestion, pollution and poor-quality housing. Urban industrial cities, such as Genoa, Liverpool and Liège, suffer from environmental decay and unemployment. In addition, several of the major EC metropolises have inner-city areas characterized by economic decline and large concentrations of people living in relative poverty (European Foundation 1986).

Little priority was allocated to regional policy by the Community in the 1950s and 1960s, even though provision was made for this in Article 2 of the Treaty of Rome. This omission stemmed partly from misplaced faith in the power of market forces to eliminate regional inequalities. It also reflected favourable national and international growth conditions in the EC6 in this period, so that unemployment levels were relatively low, especially in comparison to the 1930s. By the late 1960s there was reduced economic growth in the EC and significant concentrations of regional unemployment were re-emerging. However, it was the prospect of the first enlargement which provided the immediate spur to action. In part, the European Regional Development Fund was designed as a means of channelling EC resources to the UK so as to help offset its anticipated net budgetary deficit. However, there was also a general awareness that regional polarization could lead to disenchantment with Community membership.

Different responses to regional polarization were possible, including a strengthening of national measures, but Armstrong (1989) argues that there were important reasons for developing a separate EC regional policy. Firstly, the Commission was in a position to improve the overall efficiency of regional policy by ensuring that resources were directed to the regions with the greatest needs in the Community as a whole. Secondly, the EC had a co-ordinating role and it could avoid conflicting policy objectives being pursued in adjacent regions in neighbouring countries. Thirdly, the EC member states had vested interests in eliminating unemployment and poverty throughout the Community; otherwise there was the danger of disenchantment with the EC in some regions. Finally, the elimination of regional disparities was important in securing the consent of all member states for additional economic and political integration measures. Two further arguments can be added to this list. International regional (and other) policy measures may be necessary to combat the increasing internationalization of capital in the Community. This is especially important in view of the hypermobility of capital and the shortened duration of product cycles, so that there is increasingly rapid investment and disinvestment in regions (Damette 1980). Similarly, EC integration has led to a growth of intra-EC trade which means that fluctuations in economic activities in one area are transmitted rapidly to other regions in other member states.

While there is a logical case for the development of Community regional policy, this has proved difficult because of national rivalries and the insistence on national sovereignty. FR Germany, which would have been the largest net contributor to the new policy, refused to agree to a larger-scale, more independent EC regional policy. Consequently, since it was first inaugurated in 1975, the ERDF has tended to complement rather than replace national regional policies. The EC provides grants for infrastructural projects, as well as for industries and services. Projects are submitted via national governments and the ERDF contributes no more than 50 per cent of the assistance provided by national agencies. As such the ERDF was much more an extension of national policies than a genuine common policy. This was reinforced by the approach to designating assisted areas. The Thomson Report had tried to define areas within the EC on the basis of common objective indicators but its recommendations were politically unacceptable. As a result, individual states were allowed to designate their own eligible areas, and each country was allocated an annual expenditure quota. Italy, followed by the UK, had the largest share of such expenditure in the 1970s.

The ERDF has been subject to constant revision (table 7.3). As a result of shifts in relative regional and national prosperity in the 1970s, the ERDF was made more selective in 1982. Several northern European states were completely excluded from assistance while resources were concentrated in Ireland, Italy, the UK and the new member, Greece. In 1985 there was a further reform. Quotas were replaced by ranges, providing minimum and maximum figures for expenditure in each country. This gave the Commission greater autonomy in policy implementation, as well as providing some resources to countries such as Belgium, which were previously excluded. These were revised in 1986 following the accession of Spain and Portugal.

Further reforms followed in 1988 to take into account the increasing problems of declining, mature industrial regions in the Community. Objective 1 (peripheral) and Objective 2 (mid-decline) regions were identified (see table 7.3). Their distribution is shown in figure 7.4. Objective 1-type regions are most important in Spain, Portugal, southern Italy, Ireland and Greece, while Objective 2 regions feature most prominently in the UK, Belgium, the Ruhr and

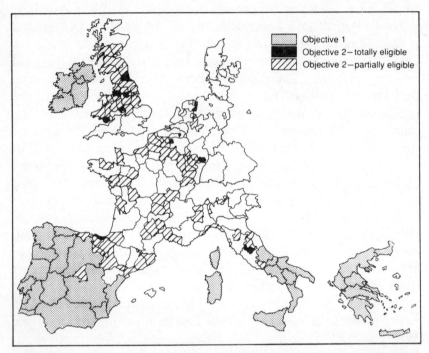

Objective 1
Objective 2—totally eligible
Objective 2—partially eligible

Figure 7.4 Areas eligible for ERDF objectives 1 and 2 assistance, 1989–93
Source: Commission of the European Communities

the Basque country. The overall allocation of resources to the EC's structural funds has also been increased. This was an important part of the 'Delors package' which aimed to ameliorate the impact of the 1992 programme on the Community's poorer regions. The targeting of the ERDF has certainly been more effective and selective; for example, the Mezzogiorno, Greece excluding Athens, Ireland, Northern Ireland and the French overseas departments secured 60 per cent of expenditure in the 1987–92 period.

The Community's structural funds were renegotiated at the December 1992 Edinburgh summit, as part of an overall agreement on budgets. Overall structural funds were to be increased to 27.4 billion ECU by 1999, while the Cohesion Fund, which was to help

Table 7.4   EC budgets for structural action, 1993–9

| | Million ECU (1992 prices) | | | |
	Structural funds and other structural operations	Cohesion Fund	Total structural actions	Total allocated to Obj. 1 regions
1993	19,777	1,500	21,277	12,328
1994	20,135	1,750	21,885	13,220
1995	21,480	2,000	23,480	14,300
1996	22,740	2,250	24,990	15,330
1997	24,026	2,500	26,526	16,396
1998	25,690	2,550	28,240	17,820
1999	27,400	2,600	30,000	19,280

Source: Commission of the European Communities

the poorer member states adjust to the challenges of the Single Market and EMU, was to increase to 2.6 billion ECU by the same date (table 7.4). Given the emphasis on social and economic cohesion, approximately 70 per cent of structural expenditure will be channelled into Objective 1 regions, which are those considered to be lagging significantly behind the remainder of the Community. In recognition of the economic and other difficulties in what had been East Germany, these German *länder* were added to the list of Objective 1 regions (figure 7.5). In addition, Hainaut (Belgium), and Merseyside and the Highlands and Islands (UK) were also added to the list of eligible regions. The maximum rate of assistance, payable in exceptional circumstances, in the four 'cohesion countries' was also raised from 75 per cent to 85 per cent.

The increased importance attached to the ERDF in recent years is symbolized by its inclusion in the Single European Act. Under Article 130 the ERDF is called upon to 'redress the principal regional imbalances in the Community through participating in the development and structural adjustment of regions lagging behind and in the conversion of declining industrial regions'. However, the operation of the ERDF is still problematical in a number of important ways. It is under-resourced and inadequate to its task but,

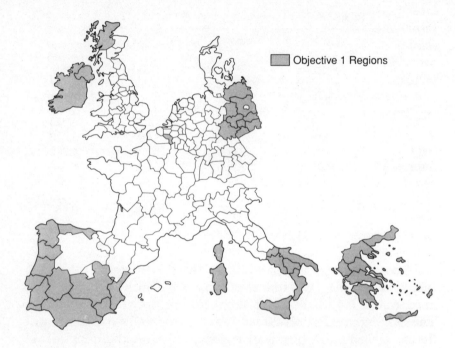

Figure 7.5   Objective 1 regions, 1993–9
Source: Commission of the European Communities

more fundamentally, it is undermined by EC expenditure in pursuit of other goals and, indeed, by the emphasis of the 1992 programme on making the Community more competitive.

The first problem of the ERDF is that of additionality. EC expenditure on regional policy is supposed to be additional to that by national governments. Yet it is probable that several member states have substituted ERDF funds for their own spending. The precise extent of this is not known but 'Considerable substitution has clearly occurred and directly undermines the ERDF's effectiveness' (Armstrong 1989, 180). Secondly, the Commission does now reserve a small proportion of the Fund to be allocated at its own discretion, rather than in accordance with national ranges or quotas. However, most projects are still filtered through national governments and resource allocation is constrained by national indicative targets (table 7.3). A genuine common policy would allow the EC to take an overview of regional problems and to devise common transnational policies to assist the areas with the greatest needs.

A third dilemma is the increasing complexity of what has been labelled 'the regional problem'. In the 1950s the main regional problems in the Community were associated with underdeveloped regions and highly localized crises in the coal, steel and textiles industries. By the 1980s regional policy was faced with a more complex economic space in the EC. There were more flexible forms of accumulation as well as the emergence of new types of problem area such as inner cities, waterfronts in decline and tourist resorts in need of restructuring (Albrechts and Swyngedouw 1989). These require both additional resources and a more flexible regional policy. A fourth problem is the smallness of the ERDF, especially when compared to the regional expenditures of some member states, and in relation to the severe difficulties faced by some problem regions. The Delors plan did provide for a major expansion of EC structural funds, including the ERDF, but it was allocated only 11 per cent of EC committed expenditure in 1989. As a result, it still could provide no more than a tiny proportion of gross fixed capital formation in most regions; Kowalski (1989) estimates that the proportion rarely surpasses 5 per cent even in the poorest EC regions. Across the EC as a whole structural expenditure accounts for less then one-third of 1 per cent of GDP (Pinder 1991, 166).

The limited expenditure on the ERDF is compounded by a further contradiction, namely the largely unintended regional effects of other EC sectoral expenditures. Capellin and Molle (1988) summarize the estimated regional effects of the EC's major sectoral expenditures (table 7.5). Agricultural policies favour the more prosperous farming areas; industrial policies have mixed effects but favour least the older industrial areas; energy policies relatively disadvantage most regions in southern Europe; transport and telecommunications policies favour the more-developed urban regions; while trade policy strongly favours the more developed

Table 7.5   Schematic view of the regional impacts of EC policies

Policy	Metropolitan		Intermediate		Agric/periph		Old industrial	
	N	*S*	*N*	*S*	*N*	*S*	*N*	*S*
Agriculture	0	0	+	0	0	–	0	0
Industry	–	0	+	+	0	0	–	–
Energy	0	–	0	–	–	–	0	0
Transport/ Telecommunication	+	0	+	–	0	–	0	0
Trade	0	0	0	0	0	0	0	0
Macro and Monetary	+	0	+	0	0	–	0	0

Source: Capellin and Molle (1988, 191)

regions. The overall impacts are difficult to assess as many of the policies have contradictory effects. However, there does seem to be a bias in favour of northern Europe and the metropolitan and intermediate regions. While this does not constitute a systematic bias, there tends to be an inverse relationship between sectoral and formal regional expenditures.

There is no intentional strategy that sectoral policies should favour the more prosperous regions of the Community. Indeed, there is a growing awareness of the potential regional implications of many EC policies. This is particularly the case with telecommunications, information technology and other high-technology industries. In consequence, there have been attempts to incorporate positive regional dimensions within sectoral policies; examples include the Special Telecommunications Action for Regions (STAR) and Science and Technology for Regional Innovation and Development in Europe (STRIDE) programmes. However, this approach is constrained by the pursuit of the EC's larger goals. The 1992 programme, with its emphasis on growth, economies of scale and productivity, could lead to greater regional inequalities (Bremm and Ache 1993; Vickerman 1990). Amin (1992, 20) believes an increase in inequalities will occur as a result of 'the exposure of less favoured regions to the free market, the exploitation of scale advantages by the more efficient players located in advanced regions and the intensification of the power and influence of TNCs whose core headquarters activities continue to be focused in major metropolitan areas'. The exception to this is the existing border regions, whose positions may be improved by the removal of barriers to trade with adjoining regions in different countries (Maillat 1990). Given that the underlying logic of this programme is to increase EC competitiveness in global markets, there are limits in the extent to which regional concerns can be accommodated. Furthermore, regional inequalities can be seen as essential prerequisites for capital accumulation, especially in terms of unequal exchange and providing labour reserves. Therefore, the fundamental logic of capitalist accumulation – of which the 1992 programme is one expression – constrains the scope for and the commitment to a regional programme. Moreover, as Planque (1993) argues in the case of France, the pressures for regional divergence will intensify if EMU is achieved, not least because the scope for state intervention-ism will be further reduced.

The Social Fund and Social Europe

Social inequality is a structural feature of capitalist development as is regional inequality. This is implicitly recognized in the Treaty of Rome with its emphasis on growth, competition and economies of scale. As such it can be seen as a manifesto for capital. Labour, and the rights of workers, barely featured in discussions prior to the Treaty or in the Treaty itself. Indeed, it only features prominently in the section on creating a common market for labour. This is not to say that social policy has been dictated in an automatic manner to serve the needs of capital. Instead, the evolution of the European Social Fund (ESF) is the outcome of the interrelationship between external pressures – especially the growth of structural unemploy-ment in the global economy – and of domestic politics in the member states.

As with regional policy, the Social Fund was predicated by a belief that market forces would eliminate major income and unemployment inequalities within the Community. Of course, it was not assumed that all social inequalities would disappear in the Community for the unequal distribution of income between capital and labour, and the profit motive, were seen as fundamental to the operation of EC markets. Nevertheless, the Treaty of Rome did allow for four main forms of social intervention (Mazey 1989b):

- The gradual achievement of the free mobility of labour, accompanied by guaranteed eligibility for social security benefits (Articles 48, 49 and 51).
- Encouragement of the exchange of young workers (Article 50).
- Equal pay for men and women (Article 119).
- The European Social Fund was to be established to assist occupational and geographical mobility (Articles 123–7).

In the late 1950s and the 1960s social policy was largely limited to measures to complete the common market for labour; that is, easing migration restrictions and securing social benefits for migrants and their families. These policies have to be seen against the background of strong economic growth and the demand for international migrant labour in northern Europe. Both these conditions eased the acceptance of EC intervention in potentially sensitive areas of national sovereignty.

By the 1970s the international economic climate was changing, with the onset of recession and the growth of unemployment. Increasing awareness of the extent of uneven development within the EC led to pressure for greater intervention in social affairs. In this climate there was also growing discrimination against emigrant workers, which threatened one of the few areas where EC social policy could claim to have been effective. In addition, the Social Democrats in FR Germany, led by Willy Brandt, were beginning to exert a greater influence on EC policy formulation; they were particularly concerned to give the Community a more 'human face'.

As a result, the 1974 Council of Ministers approved a new Social Action Programme involving forty different measures. Together with subsequent initiatives, this represented a significant shift in social policy. The earlier emphasis on assisting migrants was replaced by a new focus on youth and long-term unemployment. Since 1984 75 per cent of European Social Fund expenditure has been allocated to measures related to these (Mazey 1989b). The other special priority groups are women, the disabled, single-parent families and migrants. These are disadavantaged groups, not least because they are subject to discrimination in labour markets and experience poor access to many other opportunities (European Foundation 1986). There has also been greater spatial selectivity in ESF expenditure, with 44 per cent of funding being channelled to the 'absolute priority' areas of Greece, Ireland, Portugal, southern Italy, Northern Ireland, Corsica and the French overseas provinces, and the poorer parts of Spain.

The creation of common markets has certainly not eliminated poverty and unemployment in the EC. Indeed, there is evidence that these became more widespread in the 1980s than in any other period in the post-war era. The Commission of the European Communities (1984) noted that: 'In the mid 1970s there were 30 million people in the EC living in poverty. Today the figure, supplemented by unemployment and the growth of single parent families, is believed to be much higher.' By the late 1980s the EC estimated that there were 44 million people living in relative poverty (50 per cent below the average disposable income in each member state). To some extent this is the inevitable outcome of structural inequalities in capitalist economies. Poverty and unemployment persist even in the more socially orientated of the market economies of Western Europe, such as Sweden. Given that expenditure on the ESF is dwarfed by the social welfare programmes of the member states, it would be unrealistic to expect the former to have more than a minor impact on the distribution, let alone on the levels of poverty and unemployment. The development of the European Social Fund has, moreover, been weakened by its vague terms of reference in the Treaty of Rome, and by the claims of member states to national sovereignty over social policy. For a brief period in the 1970s, it was conceivable that the EC could develop a strong and effective ESF as part of a programme for a Social Europe. However, this:

was conceived at a time when social democracy was fashionable and national economies were buoyant. The recessions of the '70s have left behind weakened economies and a changed political climate. In short, the vision of a Social Europe has vanished. (Mazey 1988, 83)

While the ESF is incapable of addressing overall structural inequalities in the EC, the Community does impact on some areas of social inequality, particularly those that are susceptible to legal intervention. One such area is the occupational segregation of, and employment discrimination against, women. Article 119 of the Treaty of Rome, which refers to 'equal pay for equal work', provides the legal basis for EC intervention, although its origins are somewhat bizarre. As Mazey (1988) emphasizes, the principal logic behind Article 119 was to ensure that competition in the EC was not distorted by the employment of women at lower wages than were paid to men for the same work in different countries. Article 119 was not applied with any conviction by either the Commission or the member states in the 1960s, and even as late as 1975 there were differences of 20–39 per cent in the gross hourly earnings of men and women (table 7.6). This stemmed from both occupational segregation and unequal pay for similar jobs.

By the 1970s there was growing pressure to improve the labour market position of women, particularly from the increasingly well-organized and vociferous women's movements in several member states. In addition, the European Court of Justice decision in the Defrenne versus the Belgian State case closed a legal loophole which allowed national governments to escape their obligations under Article 119 (Mazey 1989a). The court ruled that Article 119 applied even if a state had no equal pay legislation. The Community's 1974 Social Action Programme also made specific reference to the needs of women in the labour market. This led directly to three important measures being approved by the Council of Ministers between 1975 and 1978: the Equal Pay Directive (75/117) guaranteeing equal pay for work of equal value; the Equal Treatment Directive (76/207) which seeks to overcome discrimination at the point of entry to the labour market; and the Social Security Directive (78/7) which

Table 7.6   Gross hourly earnings: % difference between men and women 1975–89

	1975	1980	1984	1985	1989
Belgium	28.5	29.8	26.2	25.5	24.9
Denmark	—	13.9	14.2	13.7	—
FR Germany	27.4	27.4	27.3	27.0	26.6
Greece	30.1	32.6	24.1	21.2	—
France	21.5	21.7	19.8	19.2	19.2
Ireland	39.1	31.3	31.7	32.1	31.4
Italy	20.3	15.9	15.5	—	—
Luxembourg	36.8	35.3	35.2	33.9	36.8
Netherlands	27.6	26.7	—	—	24.1
Portugal					30.6
Spain					
UK	32.1	30.4	30.5	—	31.2

Source: Mazey (1989a, 5); Commission of the European Communities (1992) *Women of Europe Supplements, no. 36*

establishes equal rights with regards to sickness and unemployment benefits.

These measures were important because they required all governments to enact legislation. The more reluctant member states were threatened with legal action which, on occasion, was resorted to. For example, a European Court decision compelled the UK government to introduce the 1986 Sex Discrimination Act to counter discrimination in small firms. The Community reinforces these measures with positive discrimination in such areas as its own expenditure on in-service training. Taken together with national measures, and with political and economic changes within individual member states (see Hudson and Williams 1989, chapter 4 on the UK), there have been some improvements in the pay and working conditions of women in the EC (table 7.6). However, there were still persistent wage inequalities in 1989. In part this is because the disadvantages of women in the labour force result from systematic disadvantages in a number of social areas such as education, household responsibilities and housing markets. As Mazey writes:

It is undoubtedly true that Community action has effectively challenged discriminatory practices in member states, which might otherwise have gone unnoticed. However, policies and directives which focus upon the economic rights of working women, do little more than scratch the surface of the problem. (1989a, 20)

While social policy was largely static during the late 1970s and the early 1980s in most areas of intervention, there was a revival in the late 1980s, spurred on by the anticipated impact of the 1992 programme. One of the major concerns was that the Single Market programme would increase the power of capital at the expense of labour. In addition, there was a fear of 'social dumping' if companies in the higher-wage economies of the EC moved to lower-wage countries in order to improve their competitiveness in the Single Market. This was of particular concern in FR Germany where, for example, some major car producers argued that reintroduction of weekend working and more flexible hours was necessary for firms to remain competitive. In reality, what was termed 'social dumping' was little different from the strategies of outward processing and subcontracting to low-cost countries which were already employed by most major Community transnationals. The pressures on working conditions, resulting from the need to maintain competitiveness, were also more significant from outside the EC (from the USA, Japan and the NICs) than within it. Another aspect of 'social dumping' was the fear that public contracts could be won by companies from other member states, and that these would import their own national lower-cost labour. To counter this fear, the Commission proposed that such companies had to abide by the social welfare and wage agreements established in the country where the work was to be carried out (Commission of the European Communities 1990).

Nevertheless, in response to these pressures, the ideas of a 'Social Europe' and of a 'European Social Space' were revived in the late 1980s. The Hannover and Rhodes European Councils both emphasized the importance of the social dimension of the 1992 programme. The communiqué from the latter Council stated:

'Completion of the Single Market cannot be regarded as an end in itself: it pursues a much wider objective, namely to ensure the maximum well-being of all, in line with the tradition of social progress which is part of Europe's history.' The President of the Commission, Jacques Delors, speaking at the European Trade Union Confederation in May 1988, also emphasized the creation of a European Social Space which ensured social and economic cohesion (Ryan 1991). This would be based on the renewal of 'social dialogue' between employers and unions at the Community level; the creation of a European Company Statute which might address the question of worker participation; and a workers' rights charter. A number of countries, including France and FR Germany, gave enthusiastic support to the idea of Social Europe. The UK viewed it as unwarranted intervention in the newly liberalized Single Market. This was not simply a case of British obstructionism, for the UK was only 'the most visible among the opponents' of labour market constraints, allowing other member states – including the Irish, Portuguese and Spanish – 'to hide behind its coat tails' (Rhodes 1991, 260).

In 1989 a draft Social Charter was produced by the Commission. It encompasses the right to freedom of movement; the right to employment and remuneration; the right to improved living and working conditions; the right to social protection; the right to freedom of association; the right to vocational training; the right of men and women to equal treatment; the right to worker participation; the right to health and safety at the workplace; the right to protection of children; the rights of elderly persons; and the rights of disabled persons (Commission of the European Community 1990). There were also specific proposals for a minimum working age and a maximum working week throughout the Community. As such, it can be seen as a direct response to the expressed fears about social dumping.

The ideas contained in the Social Charter were an important issue in the run-up to the Maastricht intergovernmental conferences. The essential logic was that the European Social Space had to be extended to keep abreast of the EC single economic space. Once again, however, the UK was implacably opposed to any such Community intervention, and it became clear that it would have to be allowed to stay out of any binding agreements. As a result, the Protocol on Social Policy in the Final Treaty on European Union notes that 'eleven member states . . . wish to continue along the path laid down in the 1989 Social Charter'. Annexed to this there is a separate agreement, which sets out what had originally been intended to be the social chapter of the Treaty. The aim of the agreement is not to introduce new measures as such but to smooth the passage of the social action programme which accompanies the Social Charter. The most important provision is the extension of qualified majority voting to cover the adoption of measures relating to working conditions, the consultation of workers, and the promotion of contractual relations between employers and employees at the European level (Gold 1992). Unanimous voting was retained on social security, the termination of contracts, the representation of and the collective defence of the interests of workers, and employment conditions for third-country nationals. The outcome is an advance in social harmonization, but it is accompanied by considerable constitutional and legal ambiguities as the result of the UK stay-out. In practice, the Commission will plan the implementation of social policy on the basis of full agreement by all twelve members, but the UK has the right to opt out. In this instance, the remaining eleven will have to resort to the Social Protocol which is technically outside of the Treaty. There are a number of contradictions inherent in this position:

• The UK is still bound by the existing articles of the EC Treaty on social policy, and these have a considerable and still largely untested range. For example the European Council extended the maternity benefits of women workers in October 1992, even though this was opposed by the UK government. As with much of the Maastricht Treaty, it seems that it largely extends existing competence.
• A qualified majority amongst the eleven under the Social Protocol is to be based on 44 votes out of 66 rather than on 54

votes out of 76. This significantly changes the balance of power for 'The worried poorer states – Spain, Portugal, Greece and Ireland – have a combined strength of only 21 votes: without Britain, they lack the muscle to block any measure the rest want to enact' (*Independent* 13 December 1991). This is likely to facilitate the introduction of anti-dumping measures which are favoured by the economically stronger member states. As such measures are viewed with increasing alarm by the poorer states, it is likely to bring to a head a simmering debate about the limits of EC interventionism in this field.

- The UK stay-out generates a number of contradictions with the aims of other EC policies, especially competition policy (because of 'unfair' UK advantages) and the notion of citizenship in the Maastricht agreement (because UK workers do not have the same rights as those elsewhere in the Community).

- If the stay-out does offer the UK competitive advantages in terms of attracting foreign investment, then the arrangement is unlikely to be sustainable in the face of opposition from the other member states. The decision by Hoover to relocate some of its production from Dijon (France) to Cambuslang (Scotland) in 1993 was taken by some critics to be the first major new instance of such social dumping. French fury was aroused because of the concessions made by the Scottish workers to attract the relocation, including limited-period contracts for new workers, constraints on the right to strike, and flexible working time and practices. In fact, the position is more complex than this, as can be seen in the case of the metal industries where Germany manufacturers are concerned about social dumping. There are considerable variations in labour costs in the EC, but this is due to a combination of wage rates *and* ancillary costs such as benefits and social security (figure 7.6). In the UK non-wage costs are only about 15 per cent of labour costs compared to 30–50 per cent in many other EC countries. This is partly because of poorer social provision in the UK, but it is also partly because the UK pays for health care through general taxation while employers have to bear a large part of the costs in many EC

Labour costs per hour per worker in DM

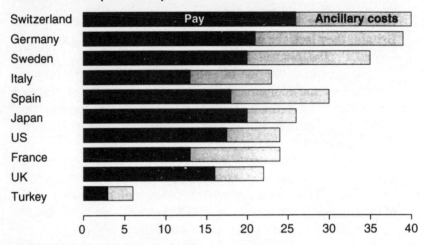

Figure 7.6   Metal industry labour costs in selected countries, 1990
Source: Metal Industry Federation

countries. At the same time, the substantially lower total labour costs in countries such as Turkey (figure 7.6) underlines the constraint that international competition places on intervention.

There certainly is no 'level playing field' within the EC in terms of working conditions. National legislation on the length of the working week is diverse, as is also the regulation of overtime working (table 7.7). Several countries have no legislation on working hours per week while, amongst those that do, the range is from 39 to 48 hours. Provision for statutory public holidays is similarly varied, and ranges from 6 to 14 days of public holidays and from 18 to 30 days of statutory leave. There is also great variety in the minimum wages paid in the EC and in the qualifying age for these. At market prices, these range from less than £250 in Spain, Portugal and Greece to more than £800 in Denmark and FR Germany (table 7.8). While the variation in real purchasing power is considerably less – from £180 to £653 – it is market rates which are important in

Table 7.7   Working hours, public holidays and statutory annual leave in the EC, 1989

(a) *Statutory regulation of working time in EC*

Country	Working week	Overtime maximum
Belgium	40 hours	65 hours per 3 months
Denmark	No legislation	Governed by collective agreement
FR Germany	48 hours	2 hours a day for up to 30 days a year on the basis of 48-hour week
Greece	5-day week, 40 hours in private sector	3 hours a day, 18 hours a week, 150 hours a year ·
Spain	40 hours	80 hours a year
France	39 hours	9 hours a week, 130 a year plus more when authorized
Ireland	48 hours	2 hours a day, 12 hours a week, 240 hours a year
Italy	48 hours .	No legislation
Luxembourg	40 hours	2 hours a day
Netherlands	48 hours	Between ½ and 3½ hours a day
Portugal	48 hours	2 hours a day, 160 a year
UK	No general legislation	No legislation

(b) *Statutory public holidays and paid annual leave in EC*

Country	Public holidays	(Paid annual leave) Statutory	(Paid annual leave) Collective agreements
Belgium	10	24 days	
Denmark	No legislation	30 days	
FR Germany	10–14	18 days	5 to 6 weeks
Greece	13	24 days	
Spain	14	30 days	
France	11	30 days	
Ireland	8	3 weeks	4 weeks
Italy	4 national + 11 others	No specific number	5 to 6 weeks
Luxembourg	10	25 days	26 to 28 days
Netherlands	6 plus one every 5 years	4 weeks	5 to 6 weeks
Portugal	12	21 to 30	
UK	No legislation	No legislation	20 to 27 days

Source: Financial Times, 12 June 1989

Table 7.8   Minimum wages in the EC, 1989

Country	Age to qualify	Level of monthly minimum wage in UK currency	
		Market	PPP
Belgium	21.5	£560.64	£495.50
Denmark	18	£826.40	£565.00
France	18	£470.30	£410.75
FR Germany	18	£209.00	£278.50
Ireland	20	£370.10– £460.45	£349.00– £435.00
Luxembourg	18	£475.90– £571.10	£498.62– £598.35
Netherlands	23	£568.20	£528.65
Portugal	20	£110.10 (agric) £116.30 (ind)	£180.65 £190.85
Spain	18	£231.65	£274.50
United Kingdom	21	£338.00– £422.50	£338.00– £422.50

Rates are per month
'Market' refers to the most recent exchange rate and 'PPP' refers to the Purchasing Power Parity exchange rate, which adjusts the exchange rate to take relative price levels into account.

Source: Minford (1989)

international economic competition. However, despite these important differences, the Commission has found it difficult to make progress in an area as sensitive as minimum wages.

The controversy over 'social dumping' may, eventually, lead to greater uniformity in working conditions, hours and terms and conditions within the EC. This could result in real improvements in labour market and employment conditions for workers in some of the poorer member states. It is ironic, however, that the driving force behind any such improvement is the concern of northern European capital to maintain competitiveness rather than social concerns with standards of living or egalitarian goals. However, even if successful, this will not refute the criticism that the EC Social Policy has been more concerned with the rights of those in work than of those out of work. At the end of 1992 there were some 16 million unemployed persons in the Community – a substantial proportion of whom formed 'a European underclass' (structurally disadvantaged). The EC has little competence, and even fewer resources, to intervene in such areas of social welfare, homelessness and poverty (Cram 1993). Furthermore, the Maastricht agreement did not seek to remedy this (Chassard and Quintin 1992). Majone (1993, 161) believes any such European welfare state is infeasible because there are so many models of welfare in Europe, 'rooted in peculiar historical and political traditions'. In the immediate future, therefore, EC social intervention is likely to be largely centred on labour market regulation.

The Environment: a Greener Europe?

The Treaty of Rome was essentially a manifesto for economic growth to be pursued via the creation of a customs union and common markets. This was hardly surprising, given the emphasis in the 1950s on achieving sustained economic expansion, little more than a decade after the end of a world war which had devastated economic structures and standards of living (Aldcroft 1980).

By the 1980s political and economic conditions had changed. A more prosperous Europe was more aware of the environmental costs of economic growth. An EC survey in 1986 showed over 55 per cent of the population in all EC countries considered that 'protection of the environment is an immediate and urgent problem' (Commission of the European Community 1987b). In Italy, Greece, Luxembourg, FR Germany and Denmark the figure surpassed 75 per cent. Issues of global significance and scale, and those which were distinctly European, were highlighted. Global warming typified the world-scale issues. It is partly caused by the release of carbon dioxide, particularly from the burning of fossil fuels in the more-developed world, coupled with depletion of the tropical rain forests. There were also issues of regional significance such as pollution of the North and Mediterranean Seas, acid rain, chemical spillages in the Rhine and other major rivers, and the nuclear fallout from the Chernobyl accident. In all these examples, it was evident that environmental disasters did not recognize national boundaries, and that Europe had a collective interest in their prevention and control.

Environmental concerns were heightened by the increasingly vociferous consumer lobby. Food production was particularly sensitive, as was illustrated by the controversy over British beef and the so-called 'mad-cow' disease in 1989–90. There was also the growth of 'green' and 'rainbow' political groups. As well as acting as a direct political force in their own right, their popularity has led to the 'greening' of the traditional European political parties. This has occurred at different rates in different countries within the EC. 'Among the Germans, seemingly sated by prosperity, ecological concerns have been seeping out from the Green movement for a decade and have become a key priority for all political parties' (*Financial Times*, 12 April 1989). The Dutch and the Danes have also been prominent in implementing measures to conserve the environment. Water quality is a major concern in the Netherlands; this is hardly surprising as it has been estimated that in the early 1980s up to 30,000 tons of salt, 3 tons of arsenic and 450 kilogrammes of mercury were dumped in the Rhine each year, even though it provides 70 per cent of Dutch drinking water (McCarthy 1989, 10). In contrast, the impact of environmental concerns on domestic politics has been considerably less in France, the UK and Southern Europe. Even so, the 1989 elections for the European Parliament revealed a considerable shift in public opinion in favour of the 'green parties' (figure 7.7). Although there are some difficulties in defining the 'green parties', they made considerable electoral gains between the 1984 and 1989 elections in several member states. Probably the most remarkable result was the 14 per cent of the vote they achieved in the UK. While the different national systems of voting did not always convert votes into equivalent proportions of elected representatives (in the UK the Greens had none), the number of seats held increased from 20 in 1984 to 39 in 1989. As a consequence, at the political level, this growing public concern involved the risk for member state

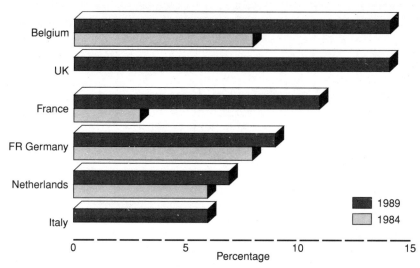

Figure 7.7   Percentage vote obtained by 'green' and 'rainbow' parties in the 1984 and 1989 European Parliament elections

governments and the Community institutions of 'losing legitimacy in case they do not take action' (Liberatore 1991, 282).

With growing public concern for and political focus on environmental questions, it was probably inevitable that these would become major issues for the EC to confront. In part this was because the distribution of environmental pollution was a trans-national issue which called for international measures. This strengthened the need for an EC policy initiative. At the same time, it constrained the role of the Community, for many environmental problems required international co-operation over much larger geographical areas. For example, one of the most notorious instances of pollution of the Rhine involved the release of chemicals into the river by the Sandoz company in Switzerland in 1987. Similarly, some of the pollution of the North Sea and the Mediterranean originates from non-EC countries. However, there were many issues, such as car exhaust emissions and food standards, on which it was possible and relevant to develop EC initiatives.

There were other reasons for EC involvement with environmental issues. Firstly, environmental pollution represented an economic cost for the member states. For example, it has been estimated that the effects of acid rain on forests, crops and buildings cost the UK some 4500 million ECU per annum, while the equivalent figure for FR Germany is 7250 million ECU. In addition, the EC's pursuit of economic growth has itself contributed to environmental pollution. For example, a large proportion of European Investment Bank funds has been channelled to potentially polluting industries such as car production, chemicals and nuclear energy. The Common Agricultural Policy, through its expenditure on modernization and on price support, has also contributed to the commercialization and intensification of farming. This, in turn, has contributed to the destruction of traditional rural landscapes and to the use of artificial fertilizers and pesticides. The ERDF has also funded potentially polluting tourist and industrial projects in the poorer regions of the EC.

Finally, it became obvious that the ability of individual member states to take measures to protect the environment was conditioned by the need to maintain competitiveness in EC markets. Environmental measures can be costly and, as was the case with the German and Dutch chemical and power-generating industries, it can seriously disadvantage domestic companies which are forced to implement them. Not surprisingly, 'The governments in Bonn and the Hague face continual pressure from local industry to demand more stringent regulations throughout the EC, so as to level the playing field' (*Financial Times*, 12 April 1989). The 1992 programme added a new twist to such arguments, for there were fears that some states would use health and environmental standards as a way of maintaining barriers to trade. This reinforced the argument for EC-wide environmental standards.

The need for an EC environmental policy was first accepted by the Council of Ministers at the 1972 European summit. This was 'constitutionally feasible' in so far as one of the aims set out by the Treaty of Rome was the improvement of the living and working conditions of EC citizens. In 1973 the First Community Action Programme on the Environment was adopted. This aimed to harmonize and co-ordinate national policies around three organizing principles: that the polluter should pay for the cost of cleaning the environment; that prevention was better than remedy; and that all decision-making should take into account environmental effects (Commission of the European Communities 1987c).

There have subsequently been Second (1977), Third (1983), Fourth (1987) and Fifth (1993) Community Action Programmes on the Environment. The Second and the Third programmes both increased the emphasis on preventive measures. Between 1973 and 1987 alone, more than 100 environmental directives were generated as a consequence of these measures (Commission of the European Communities 1987c). For example, the EC issued directives on drinking water quality in 1975 and 1980 and on bathing-water standards in 1975. In addition, there have been directives on lead and sulphur dioxide air pollution, on car exhaust emissions, on aircraft noise levels and on waste management. Of particular importance was the 1985 directive which made it obligatory to undertake environmental impact assessments prior to the approval of any large-scale industrial or infrastructural projects.

The Fourth Community Action Programme had four main emphases. Firstly, it stressed the need to integrate environmental policies more fully with other EC sectoral policies. For example, environmental goals were to be built into ERDF grants and into CAP expenditure plans. Secondly, higher and stricter environmental standards were to be set in the context of the completion of the Single Market; this was both to guarantee the level playing field and to ensure that the growth and competitiveness aims of the 1992 programme did not swamp environmental concerns. As Williams writes: 'with 1992 in mind it is critical that there is no great variation in pollution control regimes which might allow pollution havens to develop and distort competition by influencing industrial location' (1990, 205). Thirdly, there was to be increased direct EC investment in improving the environment, such as ERDF grants to clean up the

Bay of Naples and the River Mersey. Fourthly, recognizing that national governments often ignored existing EC directives, greater attention was to be paid to implementation.

It is again evident that the urgency with which non-economic policies are pursued increases markedly when they are linked to essentially economic goals such as creating a 'level playing field' for the Single Market. For similar reasons, environmental policy is often presented as being essentially a long-term *economic* policy. The Commission of the European Community (1987c, 21) has stressed this link: 'Protection of the environment is not a mere policy option. Its integration into overall economic planning is an absolute necessity. Strict environmental protection can be regarded as no less than properly conceived long-term economic policy.'

Environmental policies have also been linked with the 1992 programme in that the Single European Act explicitly defines the rights of EC institutions to legislate on environmental issues. The goals of EC policy are to conserve the environment, to protect health standards, and to ensure a rational utilization of resources. Article 130 of the SEA also provides for decisions on some environmental issues to be taken by qualified majority rather than by unanimous voting (Lodge 1989a). In practice, however, there is great scope for member states to insist on unanimous voting. The inevitable consequence of this would be a continuance of 'lowest common-denominator decision-making', which is likely to handicap policy formulation and implementation.

The Treaty on European Union agreed at Maastricht reaffirmed the environmental objectives of the Community and elevated these from the status of 'actions' to 'policies'. Community policies are to take account of: scientific data; environmental conditions in the regions; benefits versus costs of action and non-action; the overall economic and social development of the Community and its regions. The Treaty also reiterates that the principal of subsidiarity applies to environmental policies while emphasizing that member states have a duty to support Community policy in this area. In addition, the new articles in the Treaty on health and consumer protection also open up possibilities for EC interventions in the environmental field.

In 1992 a Fifth Action Programme for the Environment, entitled *Towards Sustainability*, was proposed for the period 1993–2000. It argues that success of both the Single Market and of EMU depend on sustainability and, towards this end, gives priority to improvements in:

- sustainable management of natural resources;
- integrated pollution control;
- reduced consumption of non-renewable energy sources;
- rational transport planning;
- urban environmental quality;
- public health and safety.

The programme targets five areas for special attention because of 'the particularly significant impacts they have or could have on the environment as a whole and because, by their nature, they have crucial roles to play in the attempt to achieve sustainable development'; these are energy, industry, transport, agriculture and tourism. In the last of these, for example, the aim is to promote a well-managed tourism, via diversification, improved quality and promoting 'environmentally friendly' behaviour by tourists.

Compared to previous programmes there are a number of significant differences between this and previous EC environment programmes. Reflecting the changing economic climate, there is more emphasis on the use of market instruments, that is that prices should reflect environmental costs. In recognition of the subsidiarity principle, there is a shift from top-down emphases to greater consultation and involvement of interested parties. The increased status allocated to the environment programme by the Treaty on European Union also means that it is able to make proposals for other policy areas. For example, it has been proposed that applications for ERDF funding should have to include environmental impact assessments.

While there has been a significant development of EC environmental policy in recent decades, it is still limited by a number of inherent contradictions. Firstly, the drive for uniformity – as one of the conditions of a Single Market – is likely to inhibit the governments of some countries from implementing even higher standards of environmental controls. This is particularly the case in the Netherlands, Germany and Denmark. Not only will it constrain the development of more stringent policies in these countries, but it will also lead to conflicts with other member states. This has already been demonstrated in the conflict between Germany and France over car exhaust standards. There is, however, a second and possibly more important conflict; this is between environmental policy and the objective of making the EC more competitive on the global stage. For example, German chemical companies already devote about one-quarter of their capital spending to environmental measures (*Financial Times* 5 March 1990). This could undermine their position in world markets if their competitors are operating within less-stringent regulations. In this sense, therefore, the EC may not be the appropriate international body for environmental regulation which directly affects production as opposed to consumption. Global agreements may be necessary in such cases, as is evident in the problem of damage to the ozone layer.

Decision-making and the Democratic Deficit: Neither Above nor Below the Nation State

The decision-making powers of the EC are riddled with ambiguities. As Wallace writes: 'The EC is neither "above" nor "below" the nation state. It is neither an established supranational authority, or a non-authoritative international regime. Instead it complements or acts alongside the nation state' (1982, 61). The ambiguities hinge on the division of powers between four main bodies: the Commission, the Council of Ministers, the European Parliament and the European Court of Justice. In practice, decision-making is largely the outcome of the dialogue between the Commission and the Council. However, the relationship between them has changed over time, while the Parliament has also increased its powers and role. As a first step it is necessary to outline the powers and responsibilities of these four bodies:

- *The Commission* is responsible for proposing new policies, for implementing policies, and for acting as a watchdog for the Treaty of Rome. As with any bureaucracy, its powers are strengthened by the fact that it is the main source of expert knowledge within the Community. At present there are seventeen commissioners from the twelve member states, a number arrived at by the need for political balancing. This is unwieldy, and leads to overlapping competences amongst the directorates.
- *The Council of Ministers* is the ultimate source of power since it is responsible for decision-making. The problem is that 'being the least supranational of the Community bodies it has provided a brake on developments' (Harrop 1989, 27). The acceptance of unanimous voting as part of the Luxembourg compromise greatly increased the powers of individual governments. It also contributed to the stagnation of decision-making in the 1970s and early 1980s, as each new policy usually had to offer some compensation for all member states. In some circumstances, majority voting could prevail, as the UK discovered when out-voted on agricultural prices in 1982. However, these were rare exceptions to the norm of unanimity. The Single European Act partly changed this balance in providing for majority decision-making in many areas relating to the Single Market. This did, however, imply a loss of sovereignty for national governments and, indeed, for individual national parliaments. In the case of majority voting, there are seventy-six votes distributed amongst the member states in relation to their sizes; Luxembourg has two while, at the other extreme, Germany, France, Italy and the UK have ten each. In practice it means that the big four on their own cannot secure a qualified majority of fifty-four, while it takes two large states and one medium-sized member state to muster a blocking minority of twenty-three.

- *The European Parliament* has mainly had a consultative role. However, the introduction of direct elections in 1979 has given it greater legitimacy and this has led to attempts to increase its powers. In the 1970s these were extended to approval of the EC budget, which gave it an important bargaining ploy in its negotiations with the other institutions. Under the Single European Act, the consultative powers of the Parliament were increased. There was also a new system of limited co-decision-making with the Council, with respect to legislation connected with the Single Market programme. The first draft of all legislation was sent to the Parliament to receive its opinion. Parliament could delay issuing its opinion and was able to use this as a bargaining ploy to negotiate changes in the legislation. The legislation or proposals then went to the Council of Ministers but returned to the Parliament for a second reading. At this stage the Parliament, by an absolute majority, could propose amendments to the Council's position or reject it. It then required a unanimous vote in the Council to override Parliament's views.

  There were 518 seats in the European Parliament and they were distributed amongst states according to their populations; this number has been increased to take into account German reunification. The big four each had eighty-one seats while, at the other extreme, Luxembourg had only six seats. The revised distribution of seats in 1994 is shown in figure 7.8. Voting rarely occurs along national lines so that, in a sense, the national allocation of seats is not critical. Instead, there are broad political groupings within the Parliament. The main political groupings after the 1989 election were the Socialists with 181 seats, followed by the European Peoples Party with 123 and the Liberals with 44. Membership of these groups is, however, fluid, as are their ideologies and their ability to act as effective 'parties' within the Parliament. Bittlestone writes:

  > while the member states of the EC share vague notions of Left and Right in politics and common ideologies such as Liberalism,

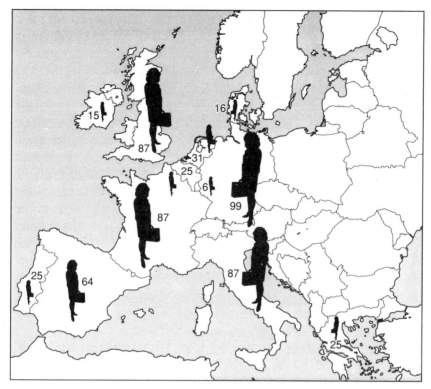

Figure 7.8   Number of seats in the European Parliament held by each
member state, 1994

Socialism, or Communism, even where the same perspective of
politics and even the same words are used, particular historical
circumstances have led to different political alignments and
perceptions of individual parties. (1989, 7)

- *The Court of Justice* acts as guardian of the Treaty of Rome and of
  EC legislation and directives. The Commission and member
  states have recourse to the Court if they consider that other
  institutions are in breach of constitutional powers. There are
  thirteen judges in the Court drawn from all the member states.

The balance of powers between these different institutions has changed during the Community's evolution (Lodge 1989b). The late 1950s and the 1960s were an era of considerable Commission discretion. It was relatively free from national government restraints in most of its actions, although the Gaullist insistence on national sovereignty meant that federalism was never really on the agenda. By the 1970s power had shifted to the European Council (the summit meetings of the heads of state). However, there was still no federalism in the governance of the EC, even though this has been proposed at several stages, most notably by the European Parliament in the 1970s and the 1980s. EC policy-making is probably best characterized as intergovernmentalism (Taylor 1983; Lodge 1989b). National governments are the central actors in the EC in what is still, essentially, a confederalist phase of integration. Their powers are institutionalized in the Council of Ministers, the European Council and in European Political Co-operation. However, the system of intergovernmentalism in the EC is different to that in other bodies of this description. This is because the legal framework of the EC allows it to intervene directly in the affairs of member states; for example, with respect to women's labour market rights, or anti-competitive practices.

The key to understanding EC policy-making lies in domestic politics, even if the reintroduction of majority voting under the Single European Act has diminished this to some degree. Domestic politics are riven by national and sectional conflicts and these filter through to EC policy-making. As a result, individual member states can be seen to take up different positions within different areas of decision-making in the Community. This provides an explanation for, amongst other things, the apparently contradictory position of Germany over agricultural policy. The German Chancellor has usually advocated reform of the CAP at the Council of Ministers, for he represents the interests of the German state, which is the largest contributor to the Community budget. However, the German Agriculture Minister, under pressure from the farm lobby, and in the context of a tradition of state interventionism in farming since the 1870s, can usually be relied upon to advocate increased expenditure on the CAP (Hendriks 1989). This is due to sectorization, whereby 'formulation of European policy in West Germany is conducted in a discrete manner according to subject matter' (Bulmer 1983, 352). In other words, the domestic political concerns of most governments with respect to most sectoral matters are transferred directly to the EC stage and do not undergo a form of Europeanization.

One area in which there has been some Europeanization is European Political Co-operation (EPC). Initially, the EC was excluded from the area of foreign policy. However, over time it became clear that 'the successful pursuit of internal policies demands an expansion of the EC's external jurisdiction and a concomitant lessening of member states' autonomy' (Lodge 1989c, 224). For example, trade and many economic interests are often indissolubly linked to other issues of diplomacy. This was eventually recognized and the 1969 Hague summit agreed that greater political co-operation was essential if a united Europe was to assume its larger responsibilities in the world. To some extent this reflected the increased self-confidence that stemmed from a decade of growth and interaction in the Community. By 1970 EPC had come into existence with the aims of co-ordinating and harmonizing the foreign policies of the member states. There are regular exchanges of information and regular meetings of foreign ministers. This was later formally institutionalized by the Single European Act. In practice, it has been difficult to harmonize diverse interests over such issues as sanctions against South Africa or relations with the Middle East, but there have been 'discreet shifts in national foreign policies towards a Euro middle ground' (Lodge 1989c, 236). The Treaty on European Union, agreed at Maastricht, has further increased EC competence in foreign policy and security matters, although it is not yet clear how this will develop in practice (see pp. 144–6).

The political contradictions of integration for the EC stem from the ambiguities of its decision-making process in the face of an expanding range of increasingly ambitious economic and social policies, especially in consequence of the 1992 programme. There had been growing pressures for greater political integration in the 1970s and early 1980s but these were sidelined by the 1985 White

Paper on the Single Market. The provision for majority voting in the Single European Act also helped to ease the political constraints on the pursuit of some of the Community's economic goals. However, several matters were left unresolved by the Single European Act. For example, many social, environmental and taxation matters are still subject to unanimous voting even though they are strongly linked to the 1992 economic programme. Any shift to greater monetary integration and or a European Central Bank would also require some transfer of powers from individual states to the centre of the EC.

In political terms the importance of the Single European Act was not to be seen in any resultant institutional changes but in that it 'reopened the debate about the inevitability of EC integration, or the survival of the nation state. This debate was intensified with the preparations for and start of, negotiations for the two new treaties on EMU and Political Union in 1991' (Kirchner 1992, 3). In the event, Maastricht promised more than it delivered in terms of political reforms. The main institutional changes were the extension of qualified majority voting to some of the Community's new areas of competence including environmental and social issues, as well as the granting of further limited co-decision-making powers to the European Parliament. Taken together, the reforms stemming from Maastricht and the Single European Act lead Kirchner (1992) to argue that decision-making in the Community can now be characterized as a cross between intergovernmentalism and co-operative federalism. The latter is to be seen in majority voting and the existence of joint tasks between national governments and Community institutions.

As Maastricht led to greater extension of EC competences in the economic, environmental and social fields, but to only limited political reforms, it fuelled the debate over the democratic deficit in the Community. This is the notion that there are substantial areas of EC activities which are not held directly accountable to either the European or national parliaments. However, the deficit can be interpreted in different ways according to the model of democracy it is compared to (Lijphart 1984). On the one hand, there is the

Westminster model which stresses the accountability of 'responsible party government' to the electorate for policies enacted. This contrasts with the consociational model which emphasizes participation in some way in policy-making, by all those affected. It is, therefore, based on coalition-building in order to achieve consensus. Decision-making in Brussels fails to match the criteria of either of these models.

The issue of decision-making was not ignored at Maastricht but, instead, it was partly redirected to the question of subsidiarity. This is the notion that decisions should be located at the lowest level at which effective decision-making is possible. The Treaty on European Union, heads of state, and members of the European Commission have all reaffirmed their commitment to subsidiarity. However, this is no more than a declaration of intent, for there is nothing in the Treaty, or in subsequent declarations such as that after the Edinburgh summit in 1992, to suggest what the appropriate levels are for particular policies. It is difficult to escape the conclusion that this is little more than a smokescreen, given that the UK, the strongest proponent of subsidiarity, has witnessed massive centralization of decision-making in the 1980s. There is anyway an argument that, given the increase in the Community's competences since the mid-1980s, the appropriate location for decision-making has shifted upwards to the level of Brussels in many fields. If this is so, resolution of the more fundamental question of the democratic deficit has only been postponed.

There is therefore pressure, resulting from the 1992 programme, the Single European Act and Maastricht for greater political union in order to provide more rapid and effective decision-making at the centre of the EC and to remedy the 'democratic deficit'. The argument is, of course, more complex than this, and there are also idealists committed to the goal of federalism as an end in itself, as well as those who seek a stronger and more coherent voice for the EC in foreign-policy circles. Against this, there are still strongly entrenched national interests and a reluctance to countenance further losses of national sovereignty. There is also considerable debate as to the form that any future political integration should take, and as to the balance of powers between the Council, the Commission and the Parliament. The contradiction for the EC is how to reconcile an increasingly harmonized economic space and integrated economic policy-making with a still largely fragmented political space.

The North–South Divide: Europe and the Third World

The EC occupies an important position in the relationships of the Third World countries because of its weight in international trade as well as the legacy of historic colonial ties. Europe – as a source of aid, as a product market and as a labour market destination – is a prime mover in the development and the dependency of much of the Third World.

Since the formation of the European Community, the trade-diversion and trade-creation effects of integration have had an impact on the trade and, therefore, the economic development of the rest of the world. This has been particularly strong in the case of agriculture; the CAP has set EC agricultural prices at above world level, thereby encouraging the expansion of Community output. The internal market for EC producers has been protected by import levies which effectively price out competitors in many sub-markets, such as beef and cereals. EC producers have also received export subsidies to increase their competitiveness in world markets, partly at the expense of Third World farmers. Industrial goods have been more favourably treated. This is because of the provisions in the Globalized System of Preferences, the Global Mediterranean Policy, agreements with individual countries and the Yaoundé and Lomé Conventions. Together these offer significant trade advantages to exports from most of the Third World, although they do exclude some of the more competitive newly industrializing countries (NICs). There are also special arrangements for particularly sensitive sectors of EC industry such as textiles, iron and steel, and cars. The first of these and agriculture are precisely the sectors in

Table 7.9 Yaoundé and Lomé Conventions,
1964–2000

Yaoundé I	1964–70
Yaoundé II	1971–75
Lomé I	1975–80
Lomé II	1980–85
Lomé III	1985–89
Lomé IV	1990–2000

which Third World countries are likely to have a competitive advantage; yet they are effectively excluded from them.

Perhaps the most significant institutional features of the relationships between the EC and the Third World are the Yaoundé and the Lomé Conventions. The first of these was signed in 1964 with the Fourth Lomé Convention coming into force in 1990 (table 7.9). Whereas the first Convention was signed by eighteen African, Caribbean and Pacific (ACP) countries, there were sixty-nine signatories to the fourth (figure 7.9). The Lomé Conventions have been controversial, for their institutionalization of relationships can be viewed in two ways:

> For the EC, the Lomé Convention represents a model of North–South cooperation, in terms of the member states' acceptance of responsibility and interest in promoting the development of the ACP states. Critics of the Convention warn that it is in reality a means for the EC to acquire secure supplies of raw materials at advantageous prices, ensure markets for production excess and maintain political influence in those areas.   (Edye and Lintner 1992, 20)

The Lomé convention offers a package of aid and trade assistance to the ACP countries. The trade concession appears to be generous, for 99.5 per cent of ACP industrial products are allowed to enter the EC tariff- and quota-free; the Community does not demand reciprocity in return. However, in practice the advantages are limited; the ACPs have relatively few industrial products, and the access on offer is, for the most part, little better than that provided by the GATT. Where

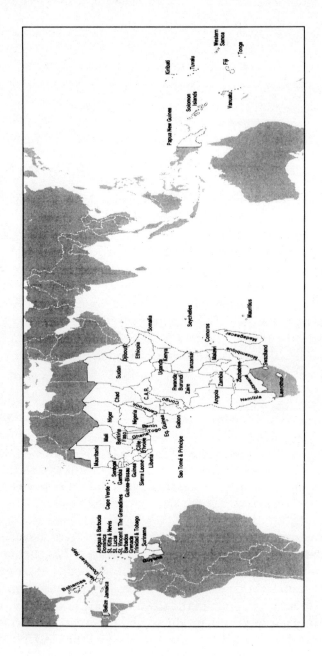

Figure 7.9   ACP signatories of Lomé IV

the ACP countries do have competitive industrial exports these are often constrained by voluntary export agreements. In addition, the EC imposes strict rules of origin on their exports; those with substantial, say, USA or Japanese components are often restricted. This, in turn, discourages foreign investment by these countries.

Aid is also provided under the Lomé convention, which is distributed via the European Development Funds. In the first five years of Lomé Four the planned expenditure is 12 billion ECU, which will be entirely in the form of grants now that soft loans have been discontinued. This is, of course, a relatively small proportion of the GDP of the EC and most aid is still provided directly by the individual member states. In addition, the amount of aid provided, relative to the populations of these countries, is on a per capita basis far smaller than is being provided to Eastern Europe (Edye and Lintner 1992); this is a striking comment on the relative priorities of the Community. Finally, there is evidence that the Lomé allocations are not necessarily well-targeted at the most needy ACP countries. Allocations reflect not only different levels of need in the receiving country, but are also influenced by historic connections (Anyadike-Danes and Anyadike-Danes 1992).

Therefore, despite the assistance provided by the EC to the Third World, in aggregate its policies may have hindered the development of the latter because of trade diversion. With the approach of the Single Market, there were concerns that the EC would become even more inward-looking. Critics argued that as the Single Market made EC industry more competitive, then there would be further trade-diversion effects as extra-EC competitors lost market shares; this was clearly recognized as a possible advantage for EC firms in the Cecchini report (see p. 106). In addition, there were concerns that common standards were being agreed as part of the drive for harmonization, without reference to the interests of non-EC exporters. The coincidence of 'Fortress Europe' with the onset of global recession and mounting debt crises in the 1980s and the 1990s all served to underline the vulnerability of the ACP countries in the world economic system.

The Third World is also linked to the EC via labour migration. In the Community as a whole there are some 12 million legal or regularized migrants, as well as an estimated additional 3 million illegal immigrants (Edye 1990). There is a distinctive geographical pattern, with Germany and France being hosts to the largest immigrant communities (see King 1993 for an overview). Thumerelle (1992) estimates that in 1989 the largest numbers of non-EC immigrants were to be found in Germany (3.5 million), France (2.1 million) and the UK (1.0 million); Belgium and the Netherlands also had large proportions of non-EC immigrants. Turkish immigrants were especially numerous in Germany (1.6 million), while immigrants from Algeria and Morocco (1.2 million) were significant in France. The cause of these migration flows is to be found in the enormous gulf in living standards between the EC and the adjoining Mediterranean region and the Third World. There are also strong demographic pressures in much of the Mediterranean region and Africa. Given the persistence of immigration into the EC, despite the recession in the 1980s and 1990s, it is clear that the push from outside the Community is greater than the pull or demand from Community labour markets, as had been the case in the 1960s. In a sense, then, emigration from the Third World reflects the failure of development to provide domestic economic opportunities; as such there is a link between emigration and the failure of the Community to provide greater assistance to this region.

Since the 1970s the member states of the Community have increased their restrictions on immigration from non-EC sources. This, however, has only fuelled illegal immigration, especially into Spain, Greece and Italy which have relatively long and difficult-to-police borders adjacent to Mediterranean and African regions of intense economic and demographic pressure. By its very nature, illegal immigration is poorly documented but there are estimates for individual countries. In Italy, for example, there were an estimated 420,000 illegal immigrants in 1990, with particularly large contingents originating from North Africa, West Africa and the Philippines (Montanari and Cortese 1993). They occupy particular segments in the labour market, especially in domestic service, street trading, agricultural harvesting and construction. Given the

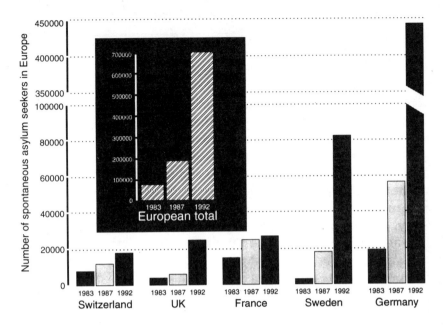

Figure 7.10 Spontaneous asylum-seekers in Europe, 1983–92
Source: United Nations High Commission for Refugees

continuous shift towards freedom of movement within the EC, there is increasing concern that there should be EC-level action to harmonize approaches to restricting immigration, protecting the rights of all immigrants, and the treatment of illegal immigrants. This has become increasingly pressing as member state governments react to the rise of far-right political movements across Europe campaigning on such issues as immigration and refugees.

The growth of refugees seeking asylum poses another challenge for the EC; whereas there were only 71,000 asylum-seekers in Europe as a whole in 1983, by 1992 this number had increased to 700,000. By far the largest number were in Germany, with some 430,000 in the latter year (figure 7.10). This reflected its relatively more generous provisions for asylum, its economic strength and its location relative to the disintegrating economies of Central and

Eastern Europe. In recognition of the differences between the member states in approaches to refugees and the increasing possibility of freedom of movement, the Community has sought to achieve policy harmonization. The most important initiative is the Schengen agreement, signed initially by France, Germany, Belgium, Luxembourg and the Netherlands, and later by Italy, Spain, and Portugal. This seeks to strengthen external border controls, harmonize visa policy, and agree common sanctions to be imposed against transport carriers bringing illegal immigrants or unauthorized refugees into the Community. In addition, the Trevi group has sought to draw up a common list of countries whose citizens require visas to enter the EC. In effect this amounts to a 'white Europe' policy, as visa-free access will only be available to the citizens of the more advanced economies. This underlines the North–South divide in migration, as in trade, and it reinforces the image of a Fortress Europe closing its gates to an impoverished Third World.

It is arguable that the incidence of wars and famines in the Third World has never been greater than in the last decade. Combined with the push of unrelenting poverty in these countries, this has led to Europe becoming one of the major centres for economic and political refugees from all parts of the globe. A case can be made on both moral and economic self-interest grounds as to why the EC should increase its assistance to the Third World via aid, trade concessions, investment, jobs and asylum. However, the growing demands on the Community have coincided with economic recession, and with competing pressures from Central and Eastern Europe. Not surprisingly, the Community has not been able to effect a significant transfer of resources to the Third World within the current narrow margins of growth. However, the Community continues to face an essential contradiction between accommodating the immediate interests of its member states and those of the poor majority of the world's population.

EIGHT

The European Community and the Future of Europe

European 'Architecture'

The European Community's economic domination of Western Europe, and especially the strength of the German economy, have increased the economic and political pressures on non-member states to redefine their relationships with the Community. In part, this is the accumulated effect of earlier enlargements, combined with the anticipated effects of the 1992 programme. The UK decision to leave EFTA and to join the EC was probably the decisive event prior to the Single Market programme. It made the EC the dominant focus of EFTA trade. By the 1980s two of the three most important export markets of all the EFTA countries were to be found in the EC (Wijkman 1989). In addition, the rules of the single market increased the pressure on all non-member states. In future, their access to EC markets would depend on a high degree of reciprocity and acceptance of EC technical, legal and environmental norms.

The EC is committed by its founding charter to make membership available to all suitable European states. In practice, the need to absorb fully the Iberian accession, to implement the 1992 programme, and to undertake other economic and political reforms, made it unlikely that the membership would be extended in the early 1990s. The question of 'appropriate members' is also far from clear, and it hangs on cultural definitions of 'European' and 'suitability'.

Adoption by EFTA (excepting Switzerland) of a package of harmonization and liberalization rules has already led to the creation of a European Economic Space. Therefore, irrespective of whether

215

Table 8.1   The EC and EFTA in 1990–2: basic comparisons

	EC	EFTA*
GNP	ECU 4,738 bn	ECU 672 bn
Population	346 m	26 m
% imports from EFTA/EC (1990)	23	61
% exports to EFTA/EC (1990)	26	58

* Norway, Sweden, Austria, Finland and Switzerland only

the EFTA states decide to become full members of the EC, they will have to accept and implement many of the Single Market measures. As Wallace writes: 'what is clear is that there is no real choice between national self-sufficiency and internationalization. The crucial choices are about the form of internationalization and over the best ways of exerting a degree of influence over the rules of the international game' (1988, 178). The choice is a difficult one. If the EFTA countries do not join the EC then they will have an uneasy relationship with it. The EC's first priority has to be effective completion of the Single Market and the Maastricht agreement, so co-operation with other countries or international associations cannot be allowed to compromise this. In addition, the EC will need to monitor the implementation of its standards and rules in the EFTA countries, but this will generate tensions. It lacks the power of inspection in EFTA countries and the European Court of Justice has no jurisdiction outside Community boundaries. On the other hand, if the EFTA countries do join the EC they will lose some of their autonomy to its collective decision-making. However, if they do not become members they will lose sovereignty anyway, through having to implement EC decisions in order to remain economically competitive.

EFTA is relatively small and its population and GNP are only equal to small proportions of those of the EC (table 8.1). Therefore, the group as a whole could be absorbed by the EC relatively painlessly. However, if the relationship between the two bodies does change it is likely to do so on a partial basis as the EFTA states have different political objectives and constraints.

- *Austria* applied for full membership in July 1989. There are compelling economic reasons for this. Two-thirds of its trade is with the EC, the schilling is informally linked to the Deutschmark, and it depends on Germany for its technology (Luetkens 1988). It has a small, relatively prosperous economy which is already well-harmonized with that of the EC so that its accession would be relatively straightforward in economic terms. The main obstacle is that the 1955 treaty which ended the Allied occupation of Austria committed it to 'permanent neutrality'. However, the relaxation in East–West relationships has diminished this obstacle. Indeed, in June 1990 the Prime Ministers of Hungary and Austria declared that their common aim was for both countries to join the EC, hence opening up the prospect of reviving some aspects of the pre-First World War Austro-Hungarian empire.
- *Norway* almost joined the EC in the early 1970s but this was narrowly rejected in a public referendum. Thereafter, Norway's special oil and fishing interests made the benefits of membership uncertain. However, by the late 1980s this was again on the agenda and Norway formally applied in 1992.
- *Sweden* had decided in 1958 that EC membership was not compatible with its neutral status. By the late 1980s the pressures of the Single Market programme, combined with the depolarization of European politics, had pushed the question back into the arena of public debate (Astrom 1988). Sweden would still find full membership politically difficult. It is sensitive to any loss of control over farm policy, given the importance of rural votes; there is concern about the implications of free movement of labour; it would be difficult to harmonize its high and progressive taxation system with that of the EC; and there are deep fears that its distinctive welfare system could be undermined. However, current reforms of the welfare state are reducing the last of these barriers. It applied for membership in 1992.
- *Finland*, historically, could not consider membership of the EC, given its neutral status and its special relationship with the

USSR. The East–West opening removed this obstacle to membership, which it applied for in 1992.

- *Switzerland*, like Sweden, is deeply committed to international neutrality. Indeed, a 1986 referendum even rejected membership of the United Nations. Having rejected membership of the European Economic Space, Switzerland has not subsequently pressed its application for full membership. Instead, Swiss transnational companies are increasing their investment within the EC as part of a defensive strategy.

The outcome of the negotiations between the EC and individual EFTA countries is likely to be critical for the future of the latter organization. If any more members leave to join the EC, EFTA may not survive as an effective trade association. The outcome of these applications is still uncertain at the time of writing. The European Council meeting at Edinburgh in 1992 agreed that enlargement negotiations with Austria, Sweden and Finland should begin in early 1993 but that they could only be concluded once the Treaty of Union had been ratified by all the member states. Norway was later included in the negotiations. The target date for full membership was set at 1 January 1995.

There are a number of difficult issues in the negotiations. Agriculture is a particularly sensitive issue with the Austrian government, with respect to Alpine farmers, and with the Scandinavian governments, with respect to Arctic farmers. They currently attract very high levels of support on social and regional grounds. In addition, retaining control of its fisheries is of intense national interest to Norway. Yet both agriculture and fisheries are sectors which already pose unresolved budgetary, political and external problems for the Community. Other important issues are the applicants' concern that the EC could erode their higher environmental standards, control of Norway's North Sea energy sources (one-half of which are currently reserved to Statoil), and the ability of the applicants to embrace incipient common foreign and defence policies given their traditions of neutrality.

There are, of course, wider implications of the proposed

enlargement of the Community from twelve to sixteen members. In the shorter term there will have to be adjustments to the numbers of Commissioners, the size of the European Parliament, and the definition of qualified majority voting. Of greater significance are the longer-term consequences for the institutional framework of the Community, that is for the 'architecture' of the EC. Will further enlargement lead to widening at the expense of deepening (greater integration)? One argument is that an increased number of members will find it more difficult to reach agreement on key issues; this will slow down decision-making in the short term, and halt further progress to European Union in the long term. More opt-out deals to accommodate the special economic and political interests of the potential new members will also increase the 'variable geometry' of the Community. This will further weaken integration in the EC. There is, however, an opposing view: the addition of new members will make it imperative that there is a reform of decision-making procedures. It has been argued that the applicants are likely to be in favour of even more majority voting in the Council (*Financial Times* 11 June 1993), especially in the areas which most concern them – social policy and the environment. All four applicants are also relatively strong and enthusiastic supporters of economic and monetary union. The prevalent view in both Finland and Sweden is that the ERM crises of 1992 and 1993, which also swamped their currencies, actually increased the case for fixed exchange rates. In the past, widening has tended to increase institutional difficulties initially but this has led to deepening in the longer term. It is difficult to escape the conclusion that the same sequence will attend the next enlargement of the Community.

In addition, the EC is faced with applications for membership from two of the smaller Mediterranean states. In 1990 Malta announced its intention to apply, as did (Greek) Cyprus. However, neither of these is likely to be as problematic as the relationship with Turkey. The 1963 Association agreement between Turkey and the EC provided a 22-year transition period for the reduction of tariffs and indicated the eventual possibility of full membership. The EC had achieved its agreed tariff reductions by 1973, with the exception of some textile products. The economic and political crises of the 1970s and the 1980s hindered Turkey but, by 1987, it was ready to apply for membership. There was opposition to the application on the basis of Turkey's low GDP per capita (less than half that of the poorest member state, Portugal), its large area and population, cultural differences stemming especially from its Islamic religion, its record on human rights and democracy, and the increased demands that would be made on the Community's structural funds. In January 1990 the Community deferred a decision on Turkey's application, influenced by the doubts expressed in the Commission's opinion, German concerns over immigration issues, and a potential Greek veto (see Balkir and Williams 1993). However, this is an issue which, in future years, is likely to test to the limits the definition of what constitutes a 'suitable European' country. The situation is further complicated by the post-1989 emergence of potential applicants in Central and Eastern Europe.

The EC and the European Homeland

In November 1989 President Gorbachev of the Soviet Union, in a now-famous speech delivered in Rome, outlined his vision of a 'common European home'. He called for the creation of a 'European legal space' within which there was 'complete uniformity in the understanding and application by all states of the norms of international law'. He added that 'we envision Europe as a commonwealth of sovereign states with a high level of equitable interdependence and easily accessible borders, open to the exchange of products, technologies and ideas, and wide-ranging contacts among people'. The East–West opening, irrespective of whether it leads to a common European homeland, was bound to have a profound impact on the EC. A European space which had been divided arbitrarily in the 1940s was now in process of, at least partial, reunification.

The reshaping of the European economy will undoubtedly have important implications for the EC, but these should not be

Table 8.2 Major features of Eastern European economies, 1991

	Bulgaria	Czecho-slovakia (former)	Hungary	Poland	Romania
Population (millions)	9.0	15.7	10.3	38.2	23.0
GDP ($bn)	7.9	33.1	30.8	78.0	27.6
Current account balance in convertible currency ($bn)	0.8	0.9	0.4	−1.3	−1.2
% share exports to EC (1989)	8	16	24	31	18
Debt service as % of exports of goods and services	22.1	11.6	32.5	5.4	2.0

Source: Economic Commission for Europe; Financial Times, 15 August 1990; World Bank (1993)

exaggerated. The total population of Poland, the former Czecho-slovakia, Hungary, Romania and Bulgaria is only 96 million (see table 8.2). Their GDP – in so far as there are reliable estimates – is only $177 billion, which is less than one-twentieth that of the EC. Poland alone accounts for 44 per cent of the total while, in contrast, Bulgaria has an estimated GDP of only $7.9 billion. Their total direct impact on the EC economy is therefore likely to be limited, at least in the short and medium term. This is reinforced by the problem of large debt burdens in countries such as Bulgaria and Hungary. In the longer term, changes in relationships with the former USSR may be of greater significance as it has a population of approximately 300 million (almost equal to that of the EC) and a GDP estimated at $1400 billion in 1989.

Despite these reservations, there are likely to be important shifts in the trading patterns of Western European states as they adjust to the new economic realities of Europe. In 1992, Germany was by far the largest exporter to Eastern Europe accounting for 52 per cent of EC trade with the region. It was followed by the other larger

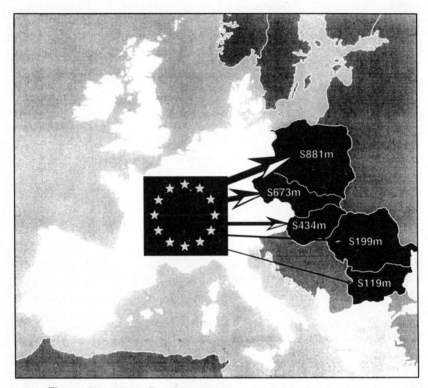

Figure 8.1 West European exports to Eastern Europe in 1992
Source: Eurostat

European economies and by those countries, such as Finland and
Austria, with special relationships with the Eastern bloc (figure 8.1).
The traditional ties of Germany with the region, together with the
undoubted strength of its own economy, will give it the lead role in
the new patterns of trade between Eastern and Western Europe.

There are constraints on the speed of change. Eastern Europe will
find it difficult to break its traditional dependence on cheap Russian
oil and on exporting often poor-quality goods to relatively protected
Eastern European markets. To counter this, the EC is providing aid
to Eastern Europe via both food aid and trade agreements. The
Community has also assumed the role of co-ordinating the special

agency established to channel investment aid to Eastern Europe, the European Reconstruction and Development Bank. However, the recovery of Eastern Europe, and its future economic relationships with the Community, are ultimately dependent on the willingness of the EC to open its markets to new sources of imports. This is highly problematic as the products in which the East has some comparative advantages are precisely those which are most sensitive in the EC: textiles, footwear, steel, coal and agriculture. In addition, the East is seeking access to EC markets at a time when the Community has set higher technical and environmental standards as part of the 1992 programme. Technologically unsophisticated and heavily polluting Eastern European firms will find it difficult to compete within these parameters.

During the course of the early 1990s the Community evolved a framework for its new relationship with Eastern and Central Europe. A series of bilateral association agreements were signed with Poland, the Czech and Slovak republics, Hungary, Romania and Bulgaria. In essence these were trade agreements. They set out ten-year timetables for the dismantling of trade barriers on industrial goods. These were 'asymmetrical' in that the Community agreed to lower tariffs and barriers earlier than the Eastern and Central European countries. There were, however, to be continuing restraints on what the EC considered to be 'sensitive sectors', such as steel, agriculture, chemicals, textiles and clothing; these, however, represented 35–45 per cent of trade. The agreements also stated that the final objective was to be full membership of the Community, but gave no dates for this. Haynes (1992, 16) comments that the 'gap between rhetoric and reality is marked' in the association agreements, while Gibb and Michalak (1993) argue that their real objective was to rule out membership in the foreseeable future.

The emerging evidence suggests that the Community has little to fear from competition in the region, at least in the short term. Since the opening to the East, exports from the EC have grown more strongly than imports into the EC from the region (figure 8.2). In 1992, compared to 1988, exports were 130 per cent higher, while

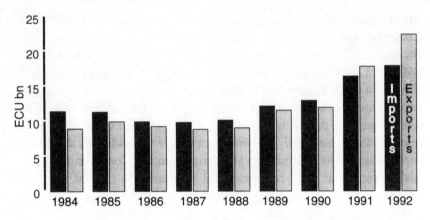

Figure 8.2   EC trade with Eastern Europe, 1984–92
Source: Eurostat

imports were up only 82 per cent. A trade deficit had also been turned into a trade surplus for the Community of 1.44 billion ECU in 1991, rising to 2.5 billion ECU in 1992. To some extent, this reflects the relative capacities of the two economies to respond to new opportunities. However, it also reflects the existence of trade barriers to what the EC terms 'sensitive sectors' but which to Eastern and Central European countries are their most competitive industries. Furthermore, the EC has been willing to increase barriers to trade in such sectors under pressure from sectional Community interest groups. For example, anti-steel-dumping measures were introduced in 1992, while imports of live animals, meat, milk and dairy products were banned in 1993, supposedly in the name of health and disease controls.

The dilemma for the Community is that it needs to balance its economic and political interests in the development of the region against narrower domestic and political pressures. Yet the two are not easily separable, as is evident in the case of migration. Ethnic conflicts, unemployment and rising aspirations have all contributed to growing pressure to emigrate from Eastern and Central Europe, principally to the EC. There are estimates that there could be as many as 40–50 million potential emigrants in the region, and although these numbers appear to be considerable overestimates (Carter et al. 1993) there is no denying the growing number of immigrants to some member states in recent years. For example, in 1990 Germany had an estimated stock of 932,000 immigrants from Eastern Europe, followed by France with 154,000. At a time of increasing surplus labour in the EC, and against the backdrop of growing far-right political influence, there is an argument that it is in the economic and political interests of Western Europe's elites to foster the development of jobs 'at home' in Central and Eastern Europe. This could be facilitated by liberalizing trade, especially in the 'sensitive sectors', and by inward investment.

Investment patterns are likely to change as new international divisions of labour are created in the European homeland. Eastern Europe represents a series of potentially attractive investment locations for EC companies. There are important new markets to be supplied and low-cost labour reserves to be tapped. In addition, major companies are attracted by the need for defensive investments or acquisitions in order to forestall similar moves by their major global competitors. Western investments in Eastern Europe are certainly not a new phenomenon, for there were already an estimated 1000 joint-ventures in the region in 1989. It is also unlikely that all Eastern European economies will be equally attractive to EC investors. For reasons of their industrial traditions and their capacity to produce a range of goods acceptable in market economies, Hungary and the former Czechoslovakia are likely to be most attractive (Dawson 1993). In contrast, Bulgaria and Romania will probably be the least attractive. This is borne out by the data for joint-ventures. Although there were an estimated 12,200 by 1992 – with 7000 in Hungary – only seventy of these were in Romania and Bulgaria (Downes and Bachtler 1992). The EC has been the main source of inward investment in Poland, Hungary and the former Czechoslovakia, with Germany accounting for more than a fifth in each case (table 8.3).

Within the first twelve months of the East–West opening, several significant investments by EC companies were announced. These include Bouygues' (France) plan to build hotels and shopping

Table 8.3   Foreign investment projects in Poland, Hungary and the former Czechoslovakia, 1991

	CSFR		Hungary		Poland	
	No. of projects	Foreign capital	No. of projects	Foreign capital	No. of projects	Foreign capital
EC	39.5	40.5	41.0	30.3	59.3	53.9
Germany	24.5	9.6	23.6	9.3	35.0	29.1
UK	3.1	5.0	4.8	5.2	5.2	3.7
Italy	3.9	0.2	4.6	5.1	4.7	5.2
France	3.1	9.9	0.9	1.0	4.6	3.8
EFTA	44.7	14.1	37.2	31.8	21.9	25.9
Austria	35.1	9.1	25.6	23.7	7.1	5.9
Switzerland	7.0	4.8	6.1	3.6	3.2	3.9
Other Europe	1.3	0.0	1.1	0.4	0.8	0.8
USA	1.8	0.0	6.5	10.8	7.6	7.9
Japan	–	–	–	–	0.2	0.3
Other	6.2	13.1	14.3	26.6	10.1	11.3
Total	100.0	100.0	100.0	100.0	100.0	100.0

Source: Downes and Bachtler (1992, 25)

centres in partnership with the Hungarian railway company and Fiat's (Italy) additional investments in its plants in Poland and the former USSR. Not surprisingly, there has been concern in the low-cost countries of the EC – such as Portugal and Greece – that they will be disadvantaged as the economic centre of Europe moves eastwards. Again, however, the pace of change is likely to be constrained. Not least, the competitive advantage of very low wages cannot be sustained for long, given the possibility of migration to neighbouring Western European economies. Even if they can be maintained, the extent to which Eastern Europe can compete with the less-developed economies in terms of low wages is limited. Furthermore, major reforms are required in currency convertibility, accounting and financial procedures and foreign investment laws before Eastern Europe will become really attractive to EC companies. At the same time, economic reforms have to be accompanied by political reforms; this will be difficult, given the lack of democratic roots in all but ex-Czechoslovakia. Fragile new regimes, long on hope but short on experience, have to deliver both economic and political 'revolutions', but in the face of a daunting world economic situation.

Given these constraints, the full impact of the opening to the East will take some time to unfold. This is not true, however, of German reunification. The two Germanies were reunited in October 1990 and single German national elections were held in December 1990. This, effectively, has brought about – in a remarkably short period of time – the fourth enlargement of the EC. There are many reasons for the speed of events, including domestic political considerations. Once the frontier had been opened, then 1:3 wage differentials between East and West were likely to lead to large-scale emigration. Unification was probably the only means of avoiding total economic collapse in East Germany.

East German membership of the EC is relatively straightforward compared to that of any of the other Eastern European countries. It had one of the stronger economies in the East, it has a shared language, and FR Germany was willing to shoulder most of the enormous costs of the transition. In addition, many West German companies had historic and/or contemporary links with plants and locations in East Germany. For example, the former headquarters of Siemens used to be in Berlin, it had trade with the East worth DM100 million and also had subcontracting links. Even so, in the short term, unification was likely to devastate the East German economy. Widespread redundancies were likely to lead to massive increases in unemployment, with rates of 20–25 per cent being widely predicted (*Financial Times* 21 May 1990). For example, one of the directors of Robotron, East Germany's largest electronics group, estimated that only 2000 of the company's 68,000 employees were producing products which could be sold in a unified market in 1990 (*Financial Times* 20 March 1990).

What will be the impact of unification on the economic geography of the EC? In the short term the impact is likely to be limited. In 1988 only 5.6 per cent of East Germany's exports were destined for West Germany, while 71 per cent of the latter's exports went to EFTA or other EC member states. It will take time for major shifts

to occur in such entrenched trading patterns. This was demonstrated by the history of integration in the EC in the 1960s and
1970s. In the longer term the economic power and dominance of
Germany will probably increase. However, this should not be
exaggerated. In 1989 FR Germany accounted for 26.1 per cent of the
GDP of the EC12, measured at market exchange rates. The second-
largest economy – France – was not very far behind with 20.5 per
cent (*Financial Times* 21 May 1990). Even had East Germany
achieved West German GDP per capita levels, then a united
Germany would still only account for 31 per cent of the GDP of the
EC. This would be approximately the same as the combined shares
of the UK and Italy. More important, but more difficult to assess, is
the extent to which Germany will be able to obtain dynamic benefits
from its location at the centre of gravity of the enlarged economic
space of the European homeland.

Another important question centres on how the opening to the
East and the reunification of Germany will affect further economic
and political union in the EC. It could be argued that the momentum
of integration will be lost as political energies are diverted to
reshaping relationships with Eastern Europe. For example, the
Belgian Prime Minister, Martens, spoke of the danger of the EC
'melting away under the warm glow of pan-Europeanism' (*Financial
Times* 14 March 1990). Federalists are concerned not only that
events in Eastern Europe will distract from the process of integration
in the EC, but that the prospect of incorporating some of these
countries as member states may lead to the Community becoming
'wider rather than deeper'. German unification, in particular, could
lessen that country's critical commitment to greater economic and
political union, not least because of the pressures on the economy
resulting from the enormous demands on German finances.
However, the German Chancellor, Kohl, had stated that unification
could only take place under a 'European roof' (*Financial Times*
2 April 1990). Consequently, federalists have been able to argue that
unification requires greater union as a means of tying a united
Germany more closely to the Community. The EC therefore seems
set for a double process of unification in the 1990s.

European Union: Unifying the Disunited

There are seven main stages of economic integration (Balassa 1961). These are:

1 a *free trade area* wherein obstacles to trade are removed;
2 a *customs union*, which is a free trade area within a common external tariff;
3 a *common market*, which is a customs union where there is also free movement of capital and labour;
4 an *economic union*, which is a common market in which there is also a high degree of co-ordination of economic policy and market regulation;
5 a *monetary union*, which is a common market with fixed exchange rates and fully convertible currencies or a single common currency;
6 *economic and monetary union*, which combines the main features of these two forms of integration;
7 *full economic union*, which is the total unification of a number of economies. As this requires a very high degree of common policy-making, it often implies that there is also political union.

In the EC the customs union was achieved before the end of 1960, at least for manufactured goods, although some non-tariff barriers remained even in this sector. The 1992 programme has effectively completed the customs union for both services and manufacturing industries. A common market for capital and labour was also established in the 1960s, although some important obstacles to the movement of professional labour and to short-term capital flows would only be eliminated by the Single Market Programme. Steps have been taken towards monetary and economic union, notably through the EMS. Although not all the member states became full members of the ERM, this was to lead to increased pressure for monetary and economic union during the 1980s and early 1990s.

At the June 1990 Dublin summit of the European Council, a decision was taken to hold two intergovernmental conferences in December 1990 on economic and monetary union and on political union. It may seem strange that further integration was being considered while the EC was still struggling through the decision-making required for the 1992 programme, and before the full impact of the Single Market could be assessed. However, as was argued earlier (see pp. 136–7), the 1992 programme itself generated much of the pressure for further reform of the Community. A single market requires greater co-ordination of economic policies, possibly involving a central European bank and a single currency. It also requires more streamlined political decision-making. In addition, it should be remembered that the 1992 programme grew out of demands for wider political reforms (see p. 105). For a while in the late 1980s, it displaced political union from the agenda, but it was probably inevitable that the latter would re-emerge. Indeed, the very success of the Single Market programme, and the greater economic integration brought about by the Single European Act, made it inevitable that there would be demands for greater political union.

While there was almost complete agreement on the need for political reform of the EC, there was considerable disagreement about its form. As the Single European Act had weakened the sovereignty of national parliaments (via majority decision-making in the Council of Ministers), there was an argument for increasing the powers of the European Parliament to compensate for this. Alternatively, the European Commission argued that there was a need for an increase in its central powers in the face of the increasing complexity of the Community. In contrast, the French government argued that there should be an increase in the powers of the Council of Ministers, allied to the introduction of more majority voting. The ultimate aims of the reformers were varied and ranged along a continuum from the Italian government's broad federalism to the UK's minimalist approach which, ideally, aimed only to create an EC macro-enterprise zone, free of economic regulation (Taylor 1989).

Also on the agenda was the possibility of a two-speed Europe. This was certainly seen as a possibility in the early 1980s when decision-making in the EC had stagnated badly. It was also an issue that hovered in the background to the UK's long-running budgetary dispute. In the mid-1980s, however, the urgency of the issue faded as the EC successfully took up the challenge of the 1992 programme. However, it re-emerged in the late 1980s as important differences surfaced both over the Exchange Rate Mechanism and the future of economic and monetary union. This was given concrete form by the June 1990 Schengen agreement. France, Germany, Belgium, the Netherlands and Luxembourg formed the free-travel zone of 'Schengenland', shifting controls on passports, visas and so on from their internal borders to their external frontier.

When the Maastricht intergovernmental conference was eventually held, there were expectations that it would mark a significant step towards economic and political union. The outcome, however, was highly uneven and while it pointed the Community in the direction of economic union, its contribution to political union was limited and in some ways contradictory. More majority decision-making was introduced into Council procedures, the European Parliament was given increased co-legislative powers, and common foreign and defence policies were mooted (see chapter Six). Against this, however, the debate before, during and after the summit brought the question of subsidiarity to the fore, not so much as a general principle of governance but as a vehicle for restating the hegemony of the nation state. There was also remarkably little in the Treaty which directly advanced the cause of federalism. Instead, the further increase in the integration of the European economic space via EMU, together with promotion of a European social space, only served to highlight the 'democratic deficit' in the European political space. In an indirect manner, therefore, Maastricht did add to the pressure for greater political union. This potential role was recognized in the Treaty on European Union agreed at Maastricht, for it contained provisions for a review of further institutional reforms in 1996. By February 1993, Delors was already indicating the likely direction of such reforms when, in a speech to the

European Parliament, he stated that the Commission 'intends to conduct a crusade for democracy'.

There is a 'democratic deficit' in the Community, but it is not clear that there is the means to rectify this. The greatest deficit lies in the procedures of the European Council. Pinder (1991, 201) writes that:

> its diplomatic methods of decision-making behind closed doors are those of the ministerial committee of an international organization, and its predominance over the Parliament in legislation and detailed control over the Commission are not suited to the house of states in a European Union.

With the exception of approving amendments to the Treaties of the European Communities, national parliaments have no power to ratify the decisions agreed in the Council during secret negotiations. The most obvious remedy to this is to increase the power of the European Parliament. However, this would entail the surrender of power by national governments, an unlikely scenario in the climate of the mid-1990s. In this sense the 'Europe of states', so stoutly defended by de Gaulle in the 1960s and by the UK in the 1980s, is likely to prevail over the European Union promoted by federalists and by states such as Belgium and the Netherlands. Whether this is sustainable in the longer term in the face of continuing economic integration and the pressure for greater regional representation at the European level is uncertain. Similarly, the likely accession of some or all of the EFTA countries will increase the pressures for a change in decision-making procedures (more majority voting, perhaps) in the Community while adding to the demand for greater transparency in democratic procedures. It is therefore likely that the 1996 review of Community institutions will be the next critical juncture in the evolution towards a European Union.

The European Community in the World Game

Until the mid-1980s the EC had little more than a supporting role to play in either of the great world games – dominance of the global economic and political systems. However, in recent years it has emerged as a major player which is challenging strongly for a place in the economic game and, to a lesser extent, the political one. A number of events have conspired to bring about this change in status. The decline of the former Soviet Union as an imperial power has rewritten the rules of the political game. At the same time, this has reshaped the political geography of Europe, bringing about the formal end of the arbitrary post-war partition. The EC has been well-placed to benefit from this in political and economic terms; this was implicitly recognized by the decision that it should co-ordinate aid for Eastern Europe and host the globally funded European Reconstruction and Development Bank. This is only one example of the growing ability of the EC to act with a single voice on the world political stage, a tendency which has been considerably strengthened by EPC (European Political Co-operation). Disunity during the Gulf and Yugoslav crises has only served to renew the demands for increased political co-ordination (see pp. 145–6).

The growing economic strength of the EC has also been important. Taken together with the difficulties in the American economy, especially its enormous trade and budgetary deficits and the fading power of the dollar, this has led to a reshaping of the global economy. American hegemony has been replaced by a tripartite relationship between the USA, Japan and the EC. In some sectors, such as agriculture, there is virtually no competition at all due to heavy protectionism. This has been one of the greatest obstacles to progress in the Uruguay round of GATT talks. However, in other sectors there is intense rivalry to dominate global markets and to establish critical leads in technology and R&D. Indeed, fear of losing out in several key high-technology sectors was one of the main stimuli to the Single Market programme. Whether this will bring about a critical increase in the competitiveness of the EC economy is an open question.

Another important question surrounds the future role of Germany in the EC. It is already the strongest economy in the Community and, in the medium term (after meeting substantial initial costs), unification is likely to strengthen its role. There are also indications in existing and emerging patterns of trade and investment that it has the greatest potential to benefit from the opening to Eastern Europe. If this is the case, and if – as seems likely – it is the largest gainer from the Single Market and from any further economic and monetary union measures, then it could overshadow the EC itself on the world stage. The post-Reagan administrations in the USA have already shown that they see Germany as the key state in the EC. Future developments may reinforce this, especially if political union lags behind economic union in the Community.

A central premise of this book is the need to interpret the evolution of the EC as the outcome of two parallel processes: globalization and a shifting series of compromises between the interests of member states. This is likely to be as true of its future as of its past evolution. There are many intriguing possibilities on the agenda in the 1990s: the United States of Europe, Social Europe, Environmental Europe, United Europe, the European Homeland, Two-tier Europe and a German-dominated European Super-economy. Whether any of these emerges, and if so in what form, will depend on the way national political interests are compromised and maximized. This, in turn, will depend on how the member states move towards political union. At the same time, there are several imponderables in the global scene. At the political level there is the whole question of the future of Russia and the other republics of the former Soviet Union. There is also the question of how allegiances will shift between the big three economic superpowers. The USA, in particular, is torn between its interests as an Atlantic power and as a Pacific power. This, in turn, is part of a larger struggle for economic supremacy between the Pacific Rim and the Atlantic Economy.

While the Single Market programme has strengthened the hand of the Community in this struggle, it is still subject to globalization processes beyond its direct influence. This was exemplified by the announcement in March 1990 that one of Europe's largest companies, Daimler-Benz, was to have wide-ranging discussions with the Japanese giant Mitsubishi about future collaboration. The fact that this occurred in the middle of a growing debate about future political and economic union in the EC should be a salutary reminder of the importance of global shifts. Similarly, the Iraq–Kuwait and the Yugoslavia crises in the 1990s also serve to underline the importance of the global context. They reveal the relative powerlessness of the EC as a diplomatic and, especially, as a military force compared to the USA.

Ultimately, there are likely to be two parallel processes of integration – at the European and at the global scale. Hence, major transnationals develop European strategies as part of larger global strategies. However, this is not an option which is open to some of the smaller transnationals. Instead there is a nested hierarchy of companies serving a range of markets from the local to the world scale. The same is true of financial markets. In the 1980s there were European and global processes of integration of stock exchanges (Corner and Tonks 1987). While Germany, the Netherlands and Italy increased their integration at the European scale, France and the UK increased their integration at the world scale. London, in particular, became increasingly linked to the Tokyo and the New York stock exchanges. The future shape of the EC economic space depends on the outcome of these parallel processes of European and global integration. However, they are not independent processes; those who are major world players will also dominate the European stage.

In the 1990s the parallel processes of Europeanization and globalization have compounded the economic difficulties faced by communities, regions and member states within the European Community. The underlying rate of unemployment has been rising in both Europe and in the G7 group of advanced economies (figure 8.3). There has been a marked slow-down in the world economy resulting in a contraction in the opportunities for EC companies. In Japan, for example, real GNP growth had averaged 4–6 per cent per annum during 1988–91, but by 1992 this had fallen to only 1 per cent. There was also the startling spectacle – almost unprecedented in recent decades – of Japanese companies implementing closures and redundancies. For example, in 1993 Nissan announced the closure of its Zama vehicle assembly plant and a 9 per cent reduction in its workforce. Japanese companies also responded to these continuing economic difficulties by further decentralizing production to lower-cost locations elsewhere in Asia and the Pacific region. The USA economy has also been struggling to achieve sustained growth in the 1990s. At the same time there has been the continuing advance of the East Asian industrial economies and the remarkable growth of the Chinese economy. Together, these various changes have resulted in depression in traditional markets for EC goods and increased competition in virtually all markets.

These difficult conditions have been paralleled by a deep structural crisis in the EC. This was symbolized by the problems faced by Daimler-Benz, which has often been presented as the flagship of the economy of Germany. In September 1993, faced with substantial deficits and declining competitivity, it announced 60,000 job losses over the period 1992–4, and an increasing propensity to shift production to low-cost locations outside Germany. More generally, a study by the European Parliament underlined the poor performance of the EC economy. It estimated that 2.5 per cent annual growth in GDP was required simply to keep unemployment levels static. In 1992 the Community achieved only a 1 per cent growth rate and witnessed a 2 million increase in unemployment. At the 1992 Edinburgh summit and elsewhere the European Council and the Commission have announced an 'EC growth package' of co-ordinated infrastructural expenditure as a short-term counter to these difficulties. At the same time a debate has opened on the long-term structural weaknesses of the EC economy in terms of R&D, labour skills and labour market flexibility.

The EC collectively faces two obstacles in its attempt to respond to these economic challenges. Firstly, can it muster sufficient consensus amongst the member states to facilitate an effective co-ordinated response in both the long and the short term? Secondly, can it persuade the other leading global economic powers to act in concert with it in an attempt to restore confidence and growth in the world economy?

This in itself is a complex issue for, as the GATT negotiations have shown, global, EC, national and regional interests are interrelated but far from coincident. There is a strong case for

a) G7 group

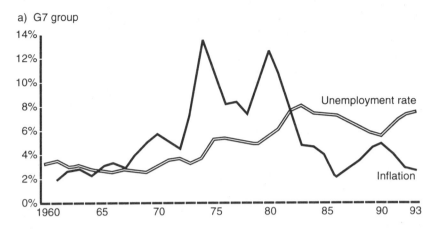

b) Four European countries (UK, Germany, France and Italy)

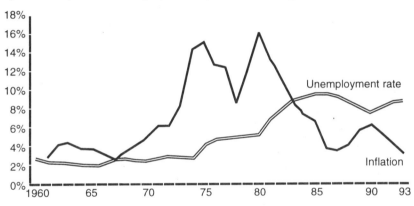

Figure 8.3   Unemployment and inflation in the G7 group compared with
Germany, Italy, France and the UK, 1960–93
Sources: Eurostat; *Financial Times*, 25 March 1993

arguing that implementation of a major GATT trade liberalizing
deal would increase global economic output, largely through trade-
creation effects. However, any such set of broad reforms would have
sharply differentiated implications for sectors and regions. French
farmers' protests against the (relatively limited) proposed restriction
of state intervention in agriculture is just one symptom of sectional

opposition to perceived disadvantages of a likely GATT deal. The EC's negotiation position is constrained by the positions taken by the other trading powers, by the conflicts within the Commission and by the pressures exerted by domestic politics. The Community is only one of the major players in the GATT negotiations, but the need to reconcile this distribution of interests has made it the key element in this particular world game.

These are not just narrow economic questions, for dependent on the outcome is the ability or willingness of the EC to assist the economic integration of the Central and Eastern European economies and, indeed, to bring about the economic convergence which is essential to EMU. It will also be easier to establish a common foreign policy and defence policy, and the Social Charter against a background of economic success. These economic challenges are not particular to the EC; what is unique is the way that the EC is responding to them through a mixture of collective Community-level and individual member-state-level initiatives. This is one final reason why the evolving shape of the EC is of such critical importance to the future of its constituent communities.

References

Ahlstrom, R.P. (1991) 'The European Community faces 1992', *Current History*, vol. 90, 374–8.

Albrechts, L. and Swyngedouw, E. (1989) 'The challenges for regional policy under a flexible regime of accumulation', in L. Albrechts, F. Moulaert, P. Roberts and E. Swyngedouw (eds), *Regional Policy at the Crossroads*, London: Jessica Kingsley and RSA.

Aldcroft, D. H. (1980) *The European Economy, 1914–1980*, London: Croom-Helm.

Allen, K., Yuill, D. and Bachtler, J. (1989) 'Requirements for an effective regional policy', in L. Albrechts, F. Moulaert, P. Roberts and E. Swyngedouw (eds), *Regional Policy at the Crossroads*, London: Jessica Kingsley and RSA.

Amin, A. (1992) *Big Firms versus the Regions in the Single European Market*, London: National Institute of Economic and Social Research, ESRC Single European Market Initiative, Working Paper no. 2.

Anyadike-Danes, M. K. and Anyadike-Danes, M. N. (1992) 'The geographic allocation of the European Development Fund under the Lomé conventions', *World Development*, vol. 20, 1647–61.

Armstrong, H. (1989) 'Community regional policy', in J. Lodge (ed.), *The European Community and the Challenge of the Future*, London: Frances Pinter.

Aström, S. (1988) 'The Nordic angle 1: Sweden's EC dilemmas', *The World Today*, vol. 44, 191–4.

Avery, G. (1987) 'Farm policy: chances for reform', *The World Today*, vol. 43, 160–65.

Baimbridge, M. and Burkitt, B. (1991) 'The Cecchini Report and the impact of 1992', *European Research*, vol. 2, 16–19.

Balassa, B. (1961) *The Theory of Etonomic Integration*, Illinois: Irwin Press.

Balkir, C. and Williams, A. M. (eds) (1993) *Turkey and Europe*, London: Frances Pinter.

Ballance, R. and Sinclair, S. (1983) *Collapse and Survival: Industrial Strategies in a Changing World*, London: George Allen and Unwin.

Barraclough, G. (1980) 'The EEC and the world economy', in D. Seers and C. Vaitsos (eds), *Integration and Unequal Development: The Experience of the EEC*, London: Macmillan.

Bittlestone, M. (1989) *Political Parties and Groups in the European Parliament*, London: Polytechnic of North London, European Dossier Series no. 5.

Blacksell, M. (1981) *Post-war Europe: A Political Geography*, London: Hutchinson.

Bos, M. and Nelson, H. (1988) 'Indirect taxation and the completion of the internal market of the EC', *Journal of Common Market Studies*, vol. 27, 27–44.

Bremm, H. J. and Ache, P. (1993) 'International changes and the Single European Market: impacts on the spatial structure of Germany', *Urban Studies*, vol. 30, 991–1007.

Bulmer, S. (1983) 'Domestic politics and European community policy-making', *Journal of Common Market Studies*, vol. 21, 349–63.

Cappellin, R. and Molle, W. (1988) 'Conclusions', in W. Molle and R. Cappellin (eds), *Regional Impact of Community Policies in Europe*, Aldershot: Avebury.

Carter, F. W., French, R. A. and Salt, J. (1993) 'International migration between East and West in Europe', *Ethnic and Racial Studies*, vol. 16, 467–91.

Castles, S., Booth, H. and Wallace T. (1984) *Here for Good: Western Europe's New Ethnic Minorities*, London: Pluto Press.

Cecchini, P. (1988) *The European Challenge – 1992: The Benefits of a Single Market*, Aldershot: Wildwood House.

Chassard, Y. and Quintin, O. (1992) 'Social protection in the European Community: towards a convergence of policies', *International Social Security Review*, vol. 45, 91–108.

Cheshire, P. (1992) *European Integration and Regional Responses*, London: National Institute of Economic and Social Research, ESRC Single European Market Initiative Working Paper no. 1.

Cheshire, P., Carbonora, G. and Hay, D. (1986) 'Problems of urban decline and growth in EEC countries: or measuring degrees of elephantness', *Urban Studies*, vol. 23, 131–49.

Church, C. (1991) *EFTA and the European Community*, London: Polytechnic of North London, European Dossier Series, no. 21.

Cini, M. (1993) *European Community Competition Policy*, London: University of North London, European Dossier Series, no. 24.

Clausse, G., Girard, J. and Rion, J.-M. (1986) 'Evolution des disparités régionales dans la communauté, 1970–1992', Luxembourg: European Investment Bank, Papers, September.

Clout, H., Blacksell, M., King, R. and Pinder, D. (1989) *Western Europe: Geographical Perspectives*, 2nd edn, Harlow: Longman.

Cobham, D. (1989) 'Strategies for monetary integration revisited', *Journal of Common Market Studies*, vol. 27, 203–18.

Collins, D. (1983) *The Operation of the European Social Fund*, Beckenham: Croom-Helm.

Commission of the European Communities (1968) *Memorandum on the Reform of Agriculture in the EEC*, Brussels: CEC, COM (68)1000.

Commission of the European Communities (1980) *Reflections on the Common Agricultural Policy*, Brussels: CEC, COM (80) 800.

Commission of the European Communities (1984) *Joint Combat Against Poverty: Background Report*, Brussels: Commission of the European Community.

Commission of the European Communities (1985a) *Perspectives for the Common Agricultural Policy*, Brussels: CEC, COM (85) 333.

Commission of the European Communities (1985b) *The European Commission's Powers of Investigation in the Enforcement of Competition Law*, Luxembourg: Commission of the European Communities.

Commission of the European Communities (1986a) *The Accession of Spain and Portugal to the European Community: A Survey*, London: Commission of the European Communities.

Commission of the European Communities (1986b) *European Unification: The Origins and Growth of the European Community*, Luxembourg: Commission of the European Communities; European Documentation 2/1986.

Commission of the European Communities (1987a) *Third Periodic Report on the Social and Economic Situation and Development of the Regions of the Community*, Brussels: Commission of the European Community, COM (87) 230 final.

Commission of the European Communities (1987b) *The European Community and Environmental Protection*, Brussels: Commission of the European Community; European File 5/87.

Commission of the European Communities (1987c) *The European*

Community and the Environment, Luxembourg: Commission of the European Communities; European Documentation 3/1987.

Commission of the European Communities (1988a) *Jean Monnet, A Grand Design for Europe*, Brussels: CEC; European Documentation 5/1988.

Commission of the European Communities (1988b) *Panorama of EC Industry*, Luxembourg: Commission of the European Communities.

Commission of the European Communities (1988c) *Europe without Frontiers: Completing the Internal Market*, Luxembourg: Commission of the European Communities.

Commission of the European Communities (1988d) *Research and Technological Development Policy*, Luxembourg: Commission of the European Communities; European Documentation 2/1988.

Commission of the European Communities (1989) *EEC Competition Policy in the Single Market*, Brussels: European Documentation 1/1989.

Commission of the European Communities (1990) *1992 – The Social Dimension*, Brussels: European Documentation 2/1990.

Commission of the European Communities (1993) *Background Report: Promoting Economic Recovery in Europe*, London: Commission of the European Communities, 16 June.

Corner, D. C. and Tonks, I. (1987) 'The impact of the internationalisation of world stock markets on the integration of EC securities markets', in M. Macmillan, D. G. Mayer and P. van Veen (eds), *European Integration and Industry*, Tilburg: Tilburg University Press.

Cowling, K. (1980) *Mergers and Economic Performance*, Cambridge: Cambridge University Press.

Cram, L. (1993) 'Calling the tune without paying the piper? Social policy regulation: the role of the Commission in European Community social policy', *Policy and Politics*, vol. 21, 135–46.

Croxford, G. J., Wise, M. and Chalkley, B. S. (1987) 'The reform of the European Regional Development Fund: a preliminary assessment', *Journal of Common Market Studies*, vol. 26, 25–38.

Damette, F. (1980) 'The regional framework of monopoly exploitation: new problems and trends', in J. Carney, R. Hudson and J. R. Lewis (eds), *Regions in Crisis*, London: Croom-Helm.

Dankert, P. (1982) 'The European Community – past, present and future', *Journal of Common Market Studies*, vol. 21, 3–18.

Darby, J. (1986) 'A new environment for public policy: Japanese manufacturing in Europe', *Western European Politics*, vol. 9, 215–34.

Dawson, A. H. (1993) *A Geography of European Integration*, London: Belhaven.

Demekas, D. G., Bartholdy, K., Gupta, S., Lipschitz, L. and Mayer, T. (1988) 'The effects of the Common Agricultural Policy of the European Community: a survey of the literature', *Journal of Common Market Studies*, vol. 27, 113–45.

Dicken, P. (1992) 'Europe 1992 and strategic change in the international automobile industry'. *Environment and Planning A*, vol. 24, 11–31.

Donges, J. B. and Schatz, K.-W. (1989) 'The Iberian countries in the EEC – risks and chances for their manufacturing industries', in G. N. Yannopoulos (ed.), *European Integration and the Iberian Economies*, Basingstoke: Macmillan.

Downes, R. and Bachtler, J. (1992) 'Foreign investment in Central and Eastern Europe', *European Business and Economic Development*, vol. 1, part 1, 22–7.

Dunning, J. H. and Pearce, R. D. (1985) *The World's Largest Industrial Enterprises, 1962–1985*, Farnborough: Gower.

Dunning, J. H. and Robson, P. (1987) 'Multinational corporate integration and regional economic integration', *Journal of Common Market Studies*, vol. 26, 103–25.

Edye, D. (1990) *1992 and the Free Movement of Labour*, London: University of North London, European Dossier Series, no. 18.

Edye, D. and Lintner, V. (1992) *The Lomé IV Convention: New Dawn or Neo-Colonialism?*, London: University of North London, European Dossier Series, no. 22.

Emerson, M., Aujean, M., Catinat, M., Goybet, P. and Jacquemin, A. (1988) *The Economics of 1992*, London: Oxford University Press.

Emminger, O. (1981) 'West Germany: Europe's driving force?, in R. Dahrendof (ed.), *Europe's Economy in Crisis*, London: Weidenfeld and Nicolson.

Eurobarometer (1987) *Eurobarometer*, no. 27, June, Brussels: Commission of the European Communities.

European Foundation for the Improvement of Living and Working Conditions (1986), *Living Conditions in Urban Areas*, Dublin: European Foundation.

Featherstone, K. (1989), 'Socialist parties in Southern Europe and the enlarged European Community', in T. Gallagher and A. M. Williams (eds), *Southern European Socialism: Parties, Elections and the Challenge of Government*, Manchester: Manchester University Press.

Fells, J. and Newman, M. (1989) *The European Community and the Superpowers*, London: Polytechnic of North London, European Dossier Series, no. 10.

Fennell, R. (1987) 'Reform of the CAP: shadow or substance?', *Journal of Common Market Studies*, vol. 26, 61–77.

Fernhout, R. (1993) ' "Europe 1993" and its refugees', *Ethnic and Racial Studies*, vol. 16, 492–505.

Gardner, B. (1987) 'The Common Agricultural Policy: the political obstacle to reform', *Political Quarterly*, vol. 58, 167–79.

Georgakopoulos, T. A. (1986) 'Greece in the European Communities: a view of the economic impact of accession', *Royal Bank of Scotland Review*, no. 150, 29–40.

Geroski, P. A. and Jacquemin, A. (1985) 'Industrial change, barriers to mobility, and European industrial policy', *Economic Policy*, no. 1, 170–218.

Gibb, R. and Michalak, W. (1993) 'The European Community and Central Europe: prospects for integration', *Geography*, vol. 78, 16–30.

Gillespie, A. (1987) 'Telecommunications and the development of Europe's less favoured regions', *Geoforum*, vol. 18, 229–39.

Ginsberg, R. H. (1989), 'US–EC relations', in J. Lodge (ed.), *The European Community and the Challenge of the Future*, London: Frances Pinter.

Gold, M. (1992) *EC Social and Labour Policy: An Overview and Update*, London: National Institute of Economic and Social Research, ESRC Single European Market Initiatives, Working Paper no. 11.

Guerrero, J. F., Gezalez, A. and Burguet, C. S. (1989) 'Spanish external trade and EEC preferences', in G. N. Yannopoulos (ed.), *European Integration and the Iberian Economies*, Basingstoke: Macmillan.

Guieu, P. and Bonnet, C. (1987) 'Completion of the internal market and indirect taxation', *Journal of Common Market Studies*, vol. 25, 209–22.

Haack, W. G. C. (1973) 'The economic effects of Britain's entry into the Common Market', *Journal of Common Market Studies*, vol. 11, 136–51.

Hallet, E. C. (1981) 'Economic convergence and divergence in the European Community: a survey of the evidence', in M. Hodges and W. Wallace (eds), *Economic Divergency in the European Community*, London: George Allen and Unwin.

Hallstein, W. (1972) *Europe in the Making*, London: George Allen and Unwin.

Hamilton, F. E. I. (1987) 'Multinational enterprises', in W. F. Lever (ed.), *Industrial Change in the United Kingdom*, Harlow: Longman.

Harrop, J. (1989) *The Political Economy of Integration in the European Community*, Aldershot: Edward Elgar.

Haynes, M. (1992) 'The rhetoric and reality of Western aid to Eastern Europe', *European Business and Economic Development*, vol. 1, part 2, 13–17.

Heath, E. (1988) 'European unity over the next ten years: from Community to Union', *International Affairs*, vol. 64, 199–207.

Hendriks, G. (1989) 'Germany and the CAP: national interests and the European Community', *International Affairs*, vol. 65, no. 1, 75–87.

Hill, B. E. (1984) *The Common Agricultural Policy: Past, Present and Future*, London: Methuen.

Hill, C. (1993) 'The capability–expectations gap, or conceptualizing Europe's international role', *Journal of Common Market Studies*, vol. 31, 305–28.

Hodges, M. (1981) 'Liberty, equality, divergency: the legacy of the Treaty of Rome?', in M. Hodges and W. Wallace (eds), *Economic Divergency in the European Community*, London: George Allen and Unwin.

Hoffman, S. (1983) 'Reflections on the nation-state in Western Europe today', in L. Tsoukalis (ed.), *The European Community: Past, Present and Future*, Oxford: Basil Blackwell.

Holland, S. (1980) *Uncommon Market*, London: Macmillan.

Hudson, R. and Williams, A. M. (1989) *Divided Britain*, London: Frances Pinter.

INSEE (1989) 'Horizon 1993: la France dans la perspective du grand marché européen', *Revue Mensuelle INSEE*, 217–18 (Paris).

Ioakimidis, P. C. (1984) 'Greece: from military dictatorship to socialism', in A. M. Williams (ed.), *Southern Europe Transformed*, London: Harper and Row.

Jacquemin, A. (1993) 'The international dimension of European competition policy', *Journal of Common Market Studies*, vol. 31, 91–102.

Jacquemin, A. and Slade, M. (1989) 'Cartels, collusion and horizontal mergers', in R. Schmalensee and R. Willig (eds), *Handbook of Industrial Organization*, Amsterdam: North-Holland.

Jenkins, R. (1984) 'Divisions over the international division of labour', *Capital and Class*, vol. 22, 28–57.

Jensen-Butler, C. (1987) 'The regional economic effects of European integration', *Geoforum*, vol. 18, 213–27.

Josling, T. E. and Mariani, A. C. (1993) 'The distributional and efficiency implications of the MacSharry proposals for reform of the CAP', *Journal of Regional Policy*, vol. 13, 27–49.

Keeble, D. (1989) 'Core–periphery disparities, recession and new regional dynamisms in the European Community', *Geography*, vol. 74, 1–11.

Kemper, N. J. and Smidt, M. de (1980) 'Foreign manufacturing establishments in the Netherlands', *Tijdschrift voor Economische en Sociale Geografie*, vol. 62, 368–82.

Kindleberger, C. P. (1967) *Europe's Post-war Growth: The Role of Labour Supply*, Cambridge, Mass.: Harvard University Press.

King, R. (1993) 'European international migration 1945–90: a statistical and geographical overview', in R. King (ed.), *Mass Migration in Europe: The Legacy and the Future*, London: Belhaven.

Kirchner, E. J. (1992) *Decision Making in the European Community: The Council Presidency and European Integration*, Manchester: Manchester University Press.

Kowalski, L. (1989) 'Major current and future regional issues in the enlarged community', in L. Albrechts, F. Moulaert, P. Roberts and E. Swyngedouw (eds), *Regional Policy at the Crossroads*, London: Jessica Kingsley and RSA.

Lebon, A. and Falchi, G. (1980) 'New developments in intra-European migration since 1974', *International Migration Review*, vol. 14, 539–73.

Lee, R. (1990) 'Making Europe: towards a geography of European integration', in M. Chisholm and D. M. Smith (eds), *Shared Space Divided Space: Essays on Conflict and Territorial Organization*, London: Unwin Hyman.

Leyshon, A. and Thrift, N. J. (1992) 'Liberalisation and consolidation: the Single European Market and the remaking of European financial capital', *Environment and Planning A*, vol. 24, 49–81.

Liberatore, A. (1991) 'Problems of transnational policy making: environmental policy in the European Community', *European Journal of Political Research*, vol. 19, 281–305.

Lijphart, A. (1984) *Democracies: Patterns of Majoritarian and Consensual Government in 21 Countries*, New Haven, Conn.: Yale University Press.

Lintner, V. (1989a) *The Common Agricultural Policy*, London: Polytechnic of North London, European Dossier Series, no. 3.

Lintner, V. (1989b) *1992: The EC Customs Union in Theory and Reality*, London: Polytechnic of North London, European Dossier Series, no. 8.

Loader, R. J. (1987) *The Structure and State of British Agriculture*, University of Reading, FMU Study no. 13.

Lodge, J. (1986) 'The European Community: compromise under domestic and international pressure', *The World Today*, vol. 42, 192–5.

Lodge, J. (1989a) 'Environment: towards a clean blue-green EC?', in J. Lodge (ed.), *The European Community and the Challenge of the Future*, London: Frances Pinter.

Lodge, J. (1989b) 'EC policymaking: institutional considerations', in J. Lodge (ed.), *The European Community and the Challenge of the Future*, London: Frances Pinter.

Lodge, J. (1989c) 'European political co-operation: towards the 1990s', in Lodge (ed.), *The European Community and the Challenge of the Future*, London: Frances Pinter.

Lodge, J. (ed.) (1989d) *The European Community and the Challenge of the Future*, London: Frances Pinter.

Ludvigen Associates (1988) *Auto-industry Opportunities and Benefits Forecast to be Generated by the Single European Market*, London: Ludvigen Associates.

Luetkens, W. L. (1988) 'Austria, EFTA and the European Community', *New European*, vol. 1, 14–17.

Lyrintzis, C. (1989) 'PASOK in power: the loss of the "third road to socialism" ', in T. Gallagher and A. M. Williams (eds), *Southern European Socialism: Parties, Elections and the Challenge of Government*, Manchester: Manchester University Press.

Madelin, A. (1988) 'A French view of the large market', *New European*, vol. 1, 8–12.

Maillat, D. (1990) 'Transborder regions between members of the EC and non-member countries', *Built Environment*, vol. 16, 38–51.

Majone, G. (1993) 'The European Community between social policy and social regulation', *Journal of Common Market Studies*, vol. 31, 153–70.

Marsh, J. S. (1989) 'The Common Agricultural Policy', in J. Lodge (ed.), *The European Community and the Challenge of the Future*, London: Frances Pinter.

Mason, C. M. and Harrison, R. T. (1990) 'Small firms: phoenix from the ashes?', in D. Pinder (ed.), *Western Europe: Challenge and Change*, London: Belhaven Press.

Mazey, S. (1988) 'European Community action on behalf of women: the limits of legislation', *Journal of Common Market Studies*, vol. 27, 63–84.

Mazey, S. (1989a) *Women and the European Community*, London: Polytechnic of North London, European Dossier Series, no. 7.

Mazey, S. (1989b) *European Community Social Policy*, London: Polytechnic of North London, European Dossier Series, no. 14.

McCarthy, E. (1989) *The European Community and the Environment*, London: Polytechnic of North London, European Dossier Series, no. 11.

McDonald, F. and Zis, G. (1989) 'The European Monetary System: towards 1992 and beyond', *Journal of Common Market Studies*, vol. 27, 183–201.

Micossi, S. (1988) 'The Single European Market: finance', *Banco Nazionale del Lavoro*, no. 165, 217–35.

Minford, M. (1989) 'Minimum wages in Europe – is Britain out of line?', *Low Pay Review*, no. 37, 20–45.

Mingione, E. (1983) 'Informalisation, restructuring and the survival strategies of the working class', *International Journal of Urban and Regional Research*, vol. 7, 311–39.

Molle, W. (1980) *Regional Disparity and Economic Development in the European Community*, Farnborough: Saxon House.

Molle, W. (1990) *The Economics of European Integration*, Aldershot: Dartmouth Publishing Company.

Molle, W. and Van Mourik, A. (1988) 'International movements of labour under conditions of economic integration: the case of Western Europe', *Journal of Common Market Studies*, vol. 26, 317–42.

Montanari, A. and Cortese, A. (1993) 'Third World immigrants in Italy', in R. King (ed.), *Mass Migration in Europe: The Legacy and the Future*, London: Belhaven.

Mowat, R. C. (1973) *Creating the European Community*, London: Blandford Press.

Mueller, D. (ed.) (1980) *The Determinants and Effects of Mergers*, Cambridge, Mass: Oelgesschlagere.

Newman, M. (1993) *The European Community: Where does the Power Lie?*, London: University of North London, European Dossier Series, no. 25.

Odell, P. R. (1976) 'The EEC energy market: structure and integration', in R. Lee and P. Ogden (eds), *Economy and Society in the EEC*, Farnborough: Saxon House.

OECD Observer (1988) 'Cutting a swathe through farm subsidies and surpluses', *OECD Observer*, no. 149, 9–11.

Osborn A. (1988) 'Greece and the EEC', *New European*, vol. 1, 13–16.

Owens, R. and Dynes, M. (1989) *1992, Britain in a Europe without Frontiers*, London: Times Books.

Padoa-Schioppa, T. (1988) 'The European Monetary System: a long term view', in F. Givazzi, S. Micossi and M. Miller (eds), *The European Monetary System*, Cambridge: Cambridge University Press.

Paine, S. (1977) 'The changing role of migrant labour in the advanced capitalist economies of Western Europe', in R. T. Griffiths (ed.), *Government, Business and Labour in European Capitalism*, London: Europotentials Press.

Pelkmans, J. and Robson, P. (1987) 'The aspirations of the White Paper', *Journal of Common Market Studies*, vol. 25, 181–92.

Pinder, J. (1964) 'The case for economic integration', *Journal of Common Market Studies*, vol. 3, 246–59.

Pinder, J. (1989) 'The Single Market: a step towards European Union', in J. Lodge (ed.), *The European Community and the Challenge of the Future*, London: Frances Pinter.

Pinder, J. (1991) *European Community: The Building of a Union*, London: Oxford University Press.

Plumb, Lord (1989) 'Building a democratic community: the role of the European Parliament', *The World Today*, vol. 45, no. 7, 112–17.

Pollard, S. (1981) *Peaceful Conquest: The Industrialisation of Europe, 1760–1970*, London: Oxford University Press.

Price, V. C. (1988) *1992: Europe's Last Chance*, London: Institute of Economic Affairs.

Regelsberger, E. (1993) 'European political cooperation', in J. Story (ed.), *The New Europe: Politics, Government and Economy since 1945*, Oxford: Basil Blackwell.

Reichenbach, H. (1980) 'A politico-economic overview', in D. Seers and C. Vaitsos (eds), *Integration and Unequal Development: The Experience of the EEC*, London: Macmillan.

Rhodes, M. (1991) 'The social dimension of the Single European Market: National versus transnational regulation', *European Journal of Political Research*, vol. 19, 245–80.

Ross, G. (1991) 'Confronting the new Europe', *New Left Review*, no. 191, 49–68.

Rothwell, R. and Dodgson, M. (1990) 'National and regional technology policies in Europe: trends and convergence', *European Research*, vol. 1, 7–13.

Rowthorn, R. and Hymer, S. (1970) 'The multinational corporation: the non-American challenge', in C. P. Kindleberger (ed.), *The International Corporation*, Boston, Mass.: Harvard University Press.

Ryan, J. (1991) 'The effects of European integration on the social policy of the European Communities', *European Research*, vol. 2, 16–21.

Sadler, D. (1992) 'Industrial policy of the European Community: strategic deficits and regional dilemmas', *Environment and Planning A*, vol. 24, 1711–30.

Salt, J. and Clout, H. D. (eds) (1976) *Migration in Post-war Europe: Geographical Essays*, Oxford: Clarendon Press.

Sauvray, J. (1984) *French Multinationals*, London: Frances Pinter.

Seers, D. (1982) 'The second enlargement in historical perspective', in D. Seers and G. Vaitsos (eds), *The Second Enlargement of the EEC*, London: Macmillan.

Servan-Schreiber, J.-J. (1968) *The American Challenge*, Harmondsworth: Pelican.

Shackleton, M. (1989) 'The budget of the European Community', in J. Lodge (ed.), *The European Community and the Challenge of the Future*, London: Frances Pinter.

Sharp, M. (1989) 'The Community and new technologies', in J. Lodge (ed.), *The European Community and the Challenge of the Future*, London: Frances Pinter.

Sharp, M. and Pavitt, K. (1993) 'Technology policy in the 1990s: old trends and new realities', *Journal of Common Market Studies*, vol. 31, 129–52.

Shearman, C. (1986) 'European collaboration in computing and telecommunications: a policy approach', *Western European Politics*, vol. 9, 147–62.

Smidt, M. de (1992) 'International investments and the European Challenge', *Environment and Planning A*, vol. 24, 83–94.

Story, J. (1981) 'Convergence at the core? The Franco-German relationship and its implications for the Community', in M. Hodges and W. Wallace (eds), *Economic Divergence in the European Community*, London: George Allen and Unwin.

Strange, S. (1987) 'The persistent myth of lost hegemony', *International Organisation*, vol. 41, 551–74.

Straubhaar, T. (1987) 'Freedom of movement of labour in a common market', *EFTA Bulletin*, vol. 28, 9–12.

Straubhaar, T. (1988) 'International labour migration within a Common Market: some aspects of EC experience', *Journal of Common Market Studies*, vol. 27, 45–61.

Strauss, R. (1983) 'Economic effects of monetary compensatory amounts', *Journal of Common Market Studies*, vol. 21, 261–81.

Swann, D. (1992) *The Single European Market and Beyond: A Study of the Wider Implications of the Single European Act*, London: Routledge.

Swinbank, A. (1993) 'CAP Reform 1992', *Journal of Common Market Studies*, vol. 31, 359–72.

Taylor, P. (1983) *The Limits of European Integration*, London: Croom-Helm.

Taylor, P. (1989) 'The new dynamics of EC integration in the 1980s', in J. Lodge (ed.), *The European Community and the Challenge of the Future*, London: Frances Pinter.

Tempini, N. (1989) *Multinational Enterprises and the European Community*, London: Polytechnic of North London, European Dossier Series, no. 9.

Thumerelle, J. (1992) 'Migrations internationales et changement géopolitique en Europe', *Annales de géographie*, no. 565, 289–318.

Tsoukalis, L. (1981) *The European Community and its Mediterranean Enlargement*, London: George Allen and Unwin.

United Nations (1988) *World Economic Survey*, 1988, New York: United Nations.

Usher, J. (1992) 'Life after Maastricht', *Parliamentary Brief*, July, 15–16.

Vickerman, R. (1990) 'Regional implications of the Single European Market', *Built Environment*, vol. 16, no. 1, 5–10.

Wallace, W. (1982) 'Europe as a confederation: the Community and the Nation-state', *Journal of Common Market Studies*, vol. 21, 57–68.

Wallace, H. (1988) 'The European Community and EFTA: one family or two?', *The World Today*, vol. 44, 177–9.

Whitelegg, J. (1988) *Transport Policy in the EEC*, London: Routledge.

Wijkman, P. M. (1989) 'Exploring the European economic space', *EFTA Bulletin*, vol. 30, no. 1, 10–15.

Williams, A. M. (1984) *Southern Europe Transformed*, London: Harper and Row.

Williams, A. M. (1987) *The Western European Economy: A Geography of Post-war Development*, London: Hutchinson.

Williams, R. (1989) 'The EC's technology policy as an engine for integration', *Government and Opposition*, vol. 24, 158–76.

Williams, R. (1990) 'Supranational environmental policy and pollution control', in D. Pinder (ed.), *Western Europe: Challenge and Change*, London: Belhaven Press.

Williamson, J. G. (1965) 'Regional inequality and the process of national development: a description of the patterns', *Economic Development and Cultural Change*, vol. 13, no. 4, 3–84.

Wise, M. and Gibb, R. (1992) *Single Market to Social Europe: The European Community in the 1990s*, Harlow, Essex: Longman.

Ziebura, G. (1982) 'Internationalisation of capital, international division of labour and the role of the European Community', *Journal of Common Market Studies*, vol. 21, 129–42.

Recommended Reading

Chapter One

Dicken, P. (1992), *Global Shift: The Internationalization of Economic Activity*, London: Paul Chapman, 2nd edn.

Story, J. (1986), *The New Europe: Politics, Government and Economy since 1945*, Oxford: Blackwell.

Williams, A. M. (1987), *The Western European Economy: A Geography of Post-War Development*, London: Hutchinson.

Chapter Two

Blacksell, M. (1981), *Post-War Europe: A Political Geography*, London: Hutchinson.

Hallstein, W. (1972), *Europe in the Making*, London: George Allen and Unwin.

Harrop, J. (1989), *The Political Economy of Integration in the European Community*, Aldershot: Edward Elgar.

Mowat, R. C. (1973), *Creating the European Community*, London: Blandford Press.

Chapter Three

Harrop, J. (1989), *The Political Economy of Integration in the European Community*, Aldershot: Edward Elgar.

Hill, B. E. (1984), *The Common Agricultural Policy: Past, Present and Future*, London: Methuen.

King, R. (1993), 'European international migration 1945–90: a statistical and geographical overview'. In R. King (ed.), *Mass Migration in Europe: The Legacy and the Future*, London: Belhaven.

Seers, D. and Vaitsos, G. (eds) (1982), *The Second Enlargement of the EEC*, London: Macmillan.

Chapter Four

Hodges, M. and Wallace, W. (eds) (1991), *Economic Divergency in the European Community*, London: George Allen and Unwin.
Seers, D. and Vaitsos, G. (eds) (1980), *Integration and Unequal Development: The Experience of the EEC*, London: Macmillan.
Taylor, P. (1983), *The Limits of European Integration*, London: Croom Helm.
Tsoukalis, L. (ed.) (1983), *The European Community: Past, Present and Future*, Oxford: Basil Blackwell.
Yannopoulos, G. N. (ed.) (1989), *European Integration and the Iberian Economies*, Basingstoke: Macmillan.

Chapter Five

Cecchini, P. (1988), *The European Challenge – 1992: The Benefits of a Single Market*, Aldershot: Wildwood House.
Lodge, J. (ed.) (1989), *The European Community and the Challenge of the Future*, London: Pinter Publishers.
Padoa-Schioppa, T. (1988), 'The European Monetary System: a long-term view'. In F. Givazzi, S. Micossi and M. Miller (eds), *The European Monetary System*, Cambridge: Cambridge University Press.
Swann, D. (1992), *The Single European Market and Beyond: A Study of the Wider Implications of the Single European Act*, London: Routledge.
Wise, M. and Gibb, R. (1992), *Single Market to Social Europe: The European Community in the 1990s*, Harlow: Longman.

Chapter Six

Pinder, J. (1991), *European Community: The Building of a Union*, Oxford: Oxford University Press.
Swann, D. (1992), *The Single European Market and Beyond: A Study of the Wider Implications of the Single European Act*, London: Routledge.
Wise, M. and Gibb, R. (1992), *Single Market to Social Europe: The European Community in the 1990s*, Harlow: Longman.

Chapter Seven

Cram, L. (1993), 'Calling the tune without paying the piper? Social policy regulation: the role of the Commission in European Community social policy', *Policy and Politics*, vol. 21, 135–46.

Jacquemin, A. (1993), 'The international dimension of European competition policy', *Journal of Common Market Studies*, vol. 31, 91–102.

Kirchner, E. J. (1992), *Decision Making in the European Community: The Council Presidency and European Integration*, Manchester: Manchester University Press.

Lodge, J. (ed.) (1989), *The European Community and the Challenge of the Future*, London: Pinter Publishers.

Molle, W. (1990), *The Economics of European Integration*, Aldershot: Dartmouth International Publishing Company.

Molle, W. and Cappellin, R. (eds) (1988), *Regional Impact of Community Policies in Europe*, Aldershot: Avebury.

Chapter Eight

Commission of the European Community (1991), *Europe 2000: Outlook for the Development of the Community's Territory*, Brussels: Commission of the European Community, Directorate General for Regional Policy.

Dawson, A. H. (1993), *A Geography of European Integration*, London: Belhaven.

Jacquemin, A. and Wright, D. (eds) (1993), *The European Challenges Post-1992: Shaping Factors and Shaping Actors*, Aldershot: Edward Elgar.

Lee R. (1990), 'Making Europe: towards a geography of European integration'. In M. Chisholm and D. M. Smith (eds), *Shared Space Divided Space: Essays on Conflict and Territorial Organization*, London: Unwin Hyman.

Index